T0334590

The Use of Mass Spectrometry Technology (MALDI-TOF) in Clinical Microbiology

THE USE OF MASS SPECTROMETRY TECHNOLOGY (MALDI-TOF) IN CLINICAL MICROBIOLOGY

Edited by

FERNANDO COBO

University Hospital Virgen de las Nieves,
Instituto Biosanitario, Granada, Spain

ACADEMIC PRESS

An imprint of Elsevier

Academic Press is an imprint of Elsevier
125 London Wall, London EC2Y 5AS, United Kingdom
525 B Street, Suite 1650, San Diego, CA 92101, United States
50 Hampshire Street, 5th Floor, Cambridge, MA 02139, United States
The Boulevard, Langford Lane, Kidlington, Oxford OX5 1GB, United Kingdom

Notices
Knowledge and best practice in this field are constantly changing. As new research and experience
broaden our understanding, changes in research methods, professional practices, or medical treatment
may become necessary.

Practitioners and researchers must always rely on their own experience and knowledge in evaluating
and using any information, methods, compounds, or experiments described herein. In using such
information or methods they should be mindful of their own safety and the safety of others, including
parties for whom they have a professional responsibility.

To the fullest extent of the law, neither the Publisher nor the authors, contributors, or editors, assume
any liability for any injury and/or damage to persons or property as a matter of products liability,
negligence or otherwise, or from any use or operation of any methods, products, instructions, or ideas
contained in the material herein.

British Library Cataloguing-in-Publication Data
A catalogue record for this book is available from the British Library

Library of Congress Cataloging-in-Publication Data
A catalog record for this book is available from the Library of Congress

ISBN: 978-0-12-814451-0

For Information on all Academic Press publications
visit our website at https://www.elsevier.com/books-and-journals

Working together
to grow libraries in
developing countries

www.elsevier.com • www.bookaid.org

Publisher: John Fedor
Acquisition Editor: Glyn Jones
Editorial Project Manager: Charlotte Rowley
Production Project Manager: Sreejith Viswanathan
Cover Designer: Mark Rogers

Typeset by MPS Limited, Chennai, India

CONTENTS

LIST OF CONTRIBUTORS

Luís Aliaga Martínez
Internal Medicine Department, University Hospital Health Campus, Granada, Spain;
Departament of Medicine, University of Granada, Granada, Spain

Marta Alonso
CIBER Enfermedades Respiratorias-CIBERES, Madrid, Spain; Servicio de
Microbiología, Hospital Universitario Donostia, Donostia, Spain

Raquel Nancy Ballesté
Clinical Laboratory Department, Hospital de Clinicas, University of the Republic,
Montevideo, Uruguay

Juan Luis Muñoz Bellido
Servicio de Microbiología y Parasitología, Complejo Asistencial Universitario de
Salamanca, Salamanca, Spain; Research Group on Clinical Microbiology and Parasitology
and Antimicrobial Resistance (IIMD-16), Instituto de Investigación Biomédica de
Salamanca (IBSAL), Universidad de Salamanca, CSIC, Complejo Asistencial Universitario
de Salamanca, Salamanca, Spain; Departamento de Ciencias Biomédicas y del
Diagnóstico, Universidad de Salamanca, Salamanca, Spain

José Manuel González Buitrago
Instituto de Investigación Biomédica de Salamanca (IBSAL), Universidad de Salamanca,
CSIC, Complejo Asistencial Universitario de Salamanca, Salamanca, Spain; Servicio de
Análisis Clínicos, Complejo Asistencial Universitario de Salamanca, Salamanca, Spain;
Departamento de Bioquímica y Biología Molecular, Universidad de Salamanca,
Salamanca, Spain

Julián Ceballos Mendiola
Department of Microbiology, Hospital Virgen de las Nieves, Granada, Spain

Fernando Cobo
Department of Microbiology, University Hospital Virgen de las Nieves, Granada, Spain

Elena Cuadros
Microbiology Hospital Universitario Virgen de las Nieves, Granada, Spain

Dra Esther Culebras
Department of Clinical Microbiology, Hospital Clínico San Carlos, Madrid, Spain

Juan de Dios Caballero
Department of Microbiology and Parasitology, Hospital Universitario Ramón y Cajal,
Madrid, Spain

Juan Carlos Rodríguez Díaz
S. Microbiología, Hospital General Universitario de Alicante-ISABIAL: Universidad
Miguel Hernández, Alicante, Spain

Sara Hernández Egido
Research Group on Clinical Microbiology and Parasitology and Antimicrobial Resistance (IIMD-16), Instituto de Investigación Biomédica de Salamanca (IBSAL), Universidad de Salamanca, CSIC, Complejo Asistencial Universitario de Salamanca, Salamanca, Spain

Maria Ercibengoa
CIBER Enfermedades Respiratorias-CIBERES, Madrid, Spain

Cristina Gómez Camarasa
Department of Microbiology, University Hospital Virgen de las Nieves, Granada, Spain

Francisco Franco-Álvarez de Luna
Microbiology Unit, UGC, Clinical Laboratory, Hospital de Riotinto, Huelva, Spain

Yannick Hoyos Mallecot
CHU Bicêtre Bacteriology-Hygiene Unit Assistance Publique Hôpitaux de Paris (AP-PH), Le Kremlin-Bicêtre, France; Associated French National Reference Center for Antibiotic Resistance: Carbapenemase-Producing Enterobacteriaceae, Le Kremlin-Bicêtre, France

Carmen Liébana-Martos
Department of Microbiology, Complejo Hospitalario de Jaén, Jaén, Spain

Oihane Martin
Department of Microbiology and Parasitology, Hospital Universitario Ramón y Cajal, Madrid, Spain

Ana Lara Oya
Department of Microbiology, Complejo Hospitalario de Jaén, Jaén, Spain

Enrique Pérez-Navarro
CAI Técnicas Biológicas, Unidad de proteómica, Departamento Microbiología, Universidad Complutense, Madrid, Spain

Zdeněk Perutka
Department of Protein Biochemistry and Proteomics, Centre of the Region Haná for Biotechnological and Agricultural Research, Faculty of Science, Palacký University, Olomouc, Czech Republic

Javier Rodriguez-Granger
Department of Microbiology, Hospital Virgen de las Nieves, Granada, Spain

María Dolores Rojo-Martín
Department of Microbiology, University Hospital Virgen de las Nieves, Granada, Spain

María Siller Ruiz
Research Group on Clinical Microbiology and Parasitology and Antimicrobial Resistance (IIMD-16), Instituto de Investigación Biomédica de Salamanca (IBSAL), Universidad de Salamanca, CSIC, Complejo Asistencial Universitario de Salamanca, Salamanca, Spain

Antonio Sampedro
Department of Microbiology, Hospital Virgen de las Nieves, Granada, Spain

Fernando Sánchez-Juanes
Instituto de Investigación Biomédica de Salamanca (IBSAL), Universidad de Salamanca, CSIC, Complejo Asistencial Universitario de Salamanca, Salamanca, Spain

Marek Šebela
Department of Protein Biochemistry and Proteomics, Centre of the Region Haná for Biotechnological and Agricultural Research, Faculty of Science, Palacký University, Olomouc, Czech Republic

Alicia Inés García Señán
Servicio de Microbiología y Parasitología, Complejo Asistencial Universitario de Salamanca, Salamanca, Spain; Research Group on Clinical Microbiology and Parasitology and Antimicrobial Resistance (IIMD-16), Instituto de Investigación Biomédica de Salamanca (IBSAL), Universidad de Salamanca, CSIC, Complejo Asistencial Universitario de Salamanca, Salamanca, Spain

Sachio Tsuchida
Department of Clinical Laboratory, Chiba University Hospital, Chiba, Japan

PREFACE

Matrix-assisted laser desorption ionization—time-of-flight (MALDI-TOF) mass spectrometry (MS) is a diagnostic tool for microbial identification and characterization based on the detection of the mass of molecules. The first description of the use of MS technology for bacterial identification was in 1975, but it took a long time for the introduction of this technique in routine microbiology. The introduction of MALDI-TOF MS in microbiological practice has supposed a great revolution due to the enormous benefits gained with the new technology. The main impact of MALDI-TOF MS in clinical diagnosis has been the rapidity in microbial identification at a relatively low cost. On the other hand, direct impact on health care and economic advantages are other characteristics that make this technique relevant in the field.

Currently, the main use of this technology in clinical microbiology is the identification or typing of bacteria from positive cultures. However, the spectrum of microorganisms which were investigated using MALDI-TOF MS permanently increased. Applications in the field of clinical virology, mycology, and microbacteriology, and the study of bacterial resistances have also been developed and are opening up new options of microbiological diagnosis.

Regarding bacteriology, MALDI-TOF MS is widely accepted as the main laboratory method for microorganism identification in the majority of laboratories, and it is possible that in the next years this technique will become the standard diagnostic tool in other microbiological areas.

In this book, experts in the field have contributed with their chapters based on their own experience from different aspects of this exciting new technology. The book is divided into 17 chapters covering all aspects of this new method. Chapters 1 and 2, Proteomics: Technology and Applications, and Basis on Mass Spectrometry: Technical Variants, respectively, involve some generalities about proteomic and the basis on MS. Chapter 3, MALDI-TOF Commercial Platforms for Microbial Identification, discusses the main commercialized MALDI-TOF MS platforms for microbial identification, and Chapters 4—6, Work Procedures in MALDI-TOF Technology; Indications, Interpretation of Results, Advantages, Disadvantages, and Limitations of MALDI-TOF; Quality Control in MALDI-TOF Techniques, respectively, cover the main issues about work procedures in MALDI-TOF MS technology, indications and interpretation of results and quality control. Chapters 7–16, Application of MALDI-TOF for Bacterial Identification; Detection of Bacterial Resistance, Biomarkers, and Virulence Factors Through MALDI-TOF Technology; Direct Identification of Pathogens From Positive Blood Cultures by MALDI-TOF Technology; Use of MALDI-TOF Techniques in the Diagnosis of Urinary Tract Pathogens; Direct Application of MALDI-TOF Mass Spectrometry to Cerebrospinal Fluid for Pathogen Identification; Application of MALDI-TOF in Clinical Virology; Use of MALDI-TOF Mass Spectrometry in Microbacterial Diagnosis; Use of MALDI-TOF Mass Spectrometry in Fungal Diagnosis; Application of MALDI-TOF in Bacterial Strain Typing and Taxonomy; Application of MALDI-TOF in Parasitology, respectively, focus on the main applications of MS technology in clinical microbiology, such as bacteria, fungi, microbacterial identification, as well as the identification of pathogens with this technique from some body fluids. Finally, Chapter 17, Future Applications of MALDI-TOF Mass Spectrometry in Clinical Microbiology, is a short review about the future applications of MALDI-TOF MS in microbiological diagnosis.

I expect that this modest contribution may help the readers to approach this exciting technology which has revolutionized the microbiological diagnosis.

I would like to acknowledge and express my sincere thanks to all contributors and colleagues from Elsevier; without their work and support the publication of this book would have been impossible.

Fernando Cobo

CHAPTER ONE

Proteomics: Technology and Applications

Raquel Nancy Ballesté
Clinical Laboratory Department, Hospital de Clinicas, University of the Republic, Montevideo, Uruguay

1.1 INTRODUCTION

Proteins are complex organic molecules, formed by amino acids arranged in long rows or polypeptide chains maintained by peptide bonds. They are vital parts of all living organisms, constituting the main components of the metabolic pathways of cells, being essential for cellular functioning and, therefore, for life [1].

Proteins have an enormous structural heterogeneity, many of which are enzymes that catalyze different chemical reactions vital for cell metabolism; others have a structural role, as cellular cytoskeleton proteins that maintain cell structure and form are fundamental in cell communication, signal transduction pathway, in the immune response, in the maintenance of cellular homeostasis, and in the cell cycle, among others.

The term proteomics was referred to in 1997 as an analogy with genomics (the study of genes). The word "proteome" is the fusion of "protein" and "genome," and was taken by Marc Wilkins in 1994. The proteome is the complete complement of proteins, including the modifications made to a particular set of proteins, produced by an organism or system [2,3].

The proteome of an organism is a highly dynamic element, because its components vary depending on the tissue, cell, or cell compartment studied, and these, in turn, can change due to alterations in their environment, such as stress situations, action of drugs, energy requirements, or their physiological state (normal or pathological). Therefore, the description of the proteome allows to have a dynamic image of all the expressed proteins, at a given moment and under certain concrete conditions of time and environment [4].

The Use of Mass Spectrometry Technology (MALDI-TOF) in Clinical Microbiology.
DOI: https://doi.org/10.1016/B978-0-12-814451-0.00001-0

The systematic study and comparison of the proteome in different metabolic and/or pathological situations allows to identify those proteins whose presence, absence, or alteration correlates with certain physiological states.

The fact that the same gene can give rise to different protein forms and these, in turn, can interact with other proteins forming protein complexes, or that the proteins present different posttranslational modifications giving rise to diverse molecular forms that can be present simultaneously, makes the proteome represent a level of complexity superior to that of the genome and that the analysis of the proteome is an even more challenging task. Unlike the genome, the proteome of an organism is dynamic, with spatiotemporal changes throughout its life cycle [5].

The Human Proteome Project has generated a map of the molecular architecture based on proteins and peptides of the human body, which is crucial to help determine the biological and molecular functions, as well as advance in the development of new applications for diagnosis, treatment, and monitoring of different pathologies [6].

1.2 DEFINITION AND IMPORTANCE OF THE PROTEOMICS

Proteomics is the large-scale study of proteins, in particular, their expression, structure, and function. Proteomics, goes beyond the mere cataloging of proteins, trying to establish, ultimately, its structure, biological activity, mode of action, cellular localization, posttranslational modifications, and interaction with other proteins or molecules [4,5].

It can also be defined as the set of techniques or technologies aimed at obtaining functional information of all proteins and is aimed at the analysis, identification, and characterization of the cellular proteome. Proteomic studies have acquired a predominant role in different applications in human pathology in recent years.

Proteomics offers highly complementary information to genomics; if we take into account that most of the biological functions are carried out by proteins, proteomics offers a new and different view of the disease. The basic difference between genomics and proteomics is that while the first one focuses in the study of the entire genetic heritage of an organism (both nuclear and extranuclear, coding or noncoding), the second focuses on the parts of the genome that are translated into proteins. These two disciplines also differ in their nature: while the genome remains relatively static, the proteome is dynamic [3,4]. The set of proteins that are expressed not only varies from one cell to another but also depends on the interactions between the genome and the environment at a specific time, so that any genome can potentially give rise to an infinite number of proteomes [4].

Currently you can find several areas of development in proteomics [4,7]:

1. Functional proteomics: study and characterization of a specific group of proteins.

2. Proteomics of expression: quantitative study of protein expression patterns between samples that differ in some variable.

3. Structural proteomics or cellular map: refers to the study of the subcellular localization of proteins and protein—protein interactions.

Despite its complexity, the rapid and remarkable development in recent years has led to its application in different pathologies or clinical proteomics, which deals with the systematic and exhaustive identification of protein patterns of disease and the application of those data to patients (study of the disease, susceptibility, prevention, selection of therapies, monitoring of treatments, etc.).

Through clinical proteomics, protein patterns can be defined to generate information of clinical value about the susceptibility, diagnosis, prognosis, and therapy of a certain disease. For this to have a real impact on health, protein patterns should be identified and selected, validated in population studies, and extrapolated to clinical practice. It is also possible to identify specific proteins associated with specific pathologies, which allow to diagnose the pathology in question or predict its evolution; these proteins are known by the generic name of biomarkers [8].

Proteomics is currently a priority line of research in the field of biology, with the growth in the number of research projects aimed at the study of proteomes in a systematic way, which has led to the emergence of new technologies on a large scale.

1.3 TECHNICAL METHODS IN PROTEOMICS

The same characteristics that give proteins their fundamental role as effector molecules of cellular function (structural and functional heterogeneity, chemical diversity, among others) also hinder their experimental analysis.

Presently there is no unique flowchart for the proteomic analysis of a sample. This is because variables, such as complexity, protein separation method, protein concentration, and stability added to the technological platform available for analysis and especially to the type of biological question intended to answer, are the basic parameters that determine the choice of a study strategy or another. There is therefore no single methodology or gold standard for the study of proteomes; consequently, proteomics research is the result of the application of a set of diverse techniques that allow the study of proteins.

The methodology for the proteomic study of a sample consists essentially of the following stages [4]:

1. Isolation and separation of proteins,
2. Analysis of the structure of the separated proteins,
3. Use of computer databases to identify the characterized proteins.

1.3.1 Isolation and separation of proteins

Biological samples are a complex mixture of proteins; usually these samples from tissues or biological fluids (blood, urine, cephalo-spinal fluid, saliva, etc.) are separated mainly by chromatographic and/or electropho retic techniques, which are robust, versatile technologies with high capacity resolution [9,10].

The most used are spectrophotometric and chromatographic systems. Electrophoresis on one-dimensional (SDS-PAGE) and two-dimensional (2-D PAGE) polyacrylamide gels is based on a separation of the proteins in function of the charge followed by a separation based on their molecular weight, capillary electrophoresis, high-performance liquid chromatography (HPLC), affinity chromatography, and ion chromatography exchange [10,11].

The types of HPLC normally used are those of normal phase, reverse phase, molecular exclusion, ion exchange, and based on bioaffinity. Normal phase, reverse phase and ion exchange HPLC separate by polarity; the one of molecular exclusion by size; and those of bioaffinity due to the different capacity of biologically active substances to form stable, specific and reversible complexes.

These isolation techniques can be used separately, altogether or in continuous flow systems, such as the multidimensional protein isolation systems technology which contains reversed phase and ion exchange in a conjugated form. Besides, another recent development in HPLC is the operation of systems of micro-HPLC and HPLC for nano-HPLC (with nano flow), presenting very high sensitivity. The use of nano-HPLC allows the separation of peptides before placing them towards the source of ionization as well as the removal of small amounts of pollutants that interfere with the analysis [10,11].

The liquid chromatography (LC) is a physical method of separation based on the distribution of the components of a mixture between two phases: a fixed or stationary and another mobile. The technology of liquid chromatography coupled mass spectrometry (LC–MS) is used for the separation of peptides of synthetic peptides, native peptides, or enzymatic proteolysis. This technology allows the separation of complex mixtures of peptides and their simultaneous analysis by mass spectrometry (MS). At present, the LC–MS system is one of the most powerful tools of the proteomic analysis based on MS.

1.3.2 Analysis of the structure of the separated proteins

Producing ions in gas phase is relatively easy for volatile compounds of low molecular weight; however, the application of MS in the field of proteins is relatively recent. This is due to the difficulty of obtaining ions of macromolecules in gas phase, because the components and proteins are compounds of low volatility and high molecular weight; therefore, the use of special ionization techniques is necessary.

The identification of proteins is performed in its vast majority by means of mass spectrometers (MSs). These instruments allow obtaining ions from organic molecules, separating and detecting them according to their mass/charge. MSs consist of three components: ionization source, mass/charge analyzer, and detector [11,12].

There are different types, the most frequently used are:

MALDI-TOF (Matrix-Assisted Laser Desorption/Ionization-Time of Flight): This is commonly used for the identification of proteins by peptide fingerprint and the study of the protein profiles of a sample and protein interactions. It is worth highlighting the development of the imaging technique, which allows a 2D visualization of the distribution of trace metals, metabolites, surface lipids, peptides, and proteins, directly in the tissues without a prior fixation or extraction [13,14].

SELDI-TOF (Surface Enhanced Laser Desorption Ionization–Time of Flight: desorption): This instrument uses the same technology as the

MALDI-TOF, incorporating the function ability of the plate (ion exchange, reverse phase, antibodies, DNA, other proteins, etc.) in which the sample is loaded. The comparison of the proteins retained in the plate among different samples can be used to search for biomarkers. This technique is an adaptation for the clinical proteomics of fluids [15].

ESI (Electro Spray Ionization): This equipment is especially useful for peptide sequencing, because it allows the ionization of macromolecules fragmented by peptide bonds (weaker bonds), thus obtaining the amino acid sequence [16].

The proteins after separation can be analyzed by MS according to different treatments. Commonly, proteins are cut out from the 2D gel where they have been isolated, digested with a protease to produce a set of peptides, and these peptides are subsequently analyzed by MS MALDI-TOF type. This technique is particularly useful for obtaining the mass spectrum of the set of peptides, known as "peptide finger-printing."

The peptides produced after a contaminant removal process are analyzed by ESI MS in combination with a triple Q (Quadrupole) or an ION TRAP, allowing the real-time study of individual peptides present in the mixture, without the need to separate them from the rest, also called as fragmentation spectrum or tandem mass spectrometry (MS/MS) [17,18].

The peptide fingerprints are characteristic of each of the proteins that allow identifying one by one in the database using bioinformatic techniques. In the same way, the MS/MS spectra are also characteristic of each of the peptides; however, none of the two techniques (peptide fingerprint or fragmentation) is of universal applicability.

In the MS, new devices are beginning to be used, which are based mainly on the hybrid combination of already known techniques. For example, devices based on the combination of a double Q and a TOF (qQ-TOF) significantly improve the resolution of the fragmentation spectra and the combination of the MALDI ionization with the qQ-TOF analyzer, or the IONTRAP allows fragmenting peptides directly once obtained from the peptide fingerprint [19,20].

The MSs called MALDI-TOF/TOF combines all the advantages of conventional MALDI-TOF spectrometry with the ability to produce peptide fragmentation spectra quickly and consistently.

There are also MSs based on the coupling of a 2D LC system with an ION TRAP or qQ-TOF analyzer. This procedure, capable of identifying thousands of peptides in a single experiment, allows analyzing automatically the components of very complex peptide mixtures. This technique (also called shotgun-proteomics) has been applied to the direct identification of whole proteomes without the separation of their components by electrophoresis [21].

Another variant of this methodology allows studies of differential expression of proteins in the proteome. This technique, known as "isotope coded affinity tag," is based on the use of protein reagents in the form of stable isotopes, which allow to determine the relative proportion of the peptides derived from the proteomes to be compared. This procedure through specific software allows the automatic and simultaneous identification of relative changes in protein concentrations and the nature of proteins that undergo differential expression.

Finally, these methodologies, together with affinity selective purification procedures, allow the analysis of modified peptides, such as phosphopeptides, from whole proteomes, which has given rise to the concept of phosphoproteome.

The Array or Protein Biochips allows the detection, characterization, and quantification of protein, as well as the study of the functional qualities of proteins and their interactions, both between them and with molecules of DNA or lipids. Unlike the nucleic acids, the proteins have neither homogeneous structure nor specific union pattern, but every protein possesses a few particular biochemical characteristics; therefore, the development of the microarrays of proteins is still before technical difficulties [22,23].

1.3.3 Utilization of computer databases to identify the characterized proteins

The interpretation of mass spectrum is the most difficult part of this methodology, as it involves the knowledge of the chemical and biochemical processes applied to protein samples, with the knowledge of bioinformatics for the search and identification of proteins in database banks, mainly because it is based on the correlation of values and arithmetic differences of atomic mass units with the molecular structure of proteins [24,25].

Proteomic analyses generate a large amount of data to analyze, which are assisted by softwares that face the spectra obtained from different databases [24]. The development of bioinformatic tools, such as the introduction of new algorithms, has been key to manipulate these collections of data. This field allows the manipulation of data on a large scale, developing programs and searching tools. These tools have been applied with great success in MS data processing in peptide fingerprint analysis (protein identification), peptide fragmentation fingerprint (identification of the peptide sequence), and de novo sequencing [24,25,26].

1.4 PROTEOMICS' ROLE IN THE CLINICAL LABORATORY

A few years ago the molecular technique and the proteomics technique were only obtained in complex laboratories. Recently we are witnessing a generalization in the use of both techniques begging to generalize in the field of diagnosis and being available in laboratories for daily practice. This, on the one hand, is a great advance for clinical laboratories, on the other hand, an enormous responsibility in the selection and correct application of mentioned methodologies. Clinical laboratories have a fundamental role in the incorporation of technology, the selection of equipment, the technique(s) to be implemented, the application of these techniques in different pathological processes, the verification and validation of the selected method, the control of quality of the preanalytical, analytical, and postanalytical processes, and the cost−benefit evaluation of the incorporation of the new procedure.

In fact, the validation in clinical practice of newly discovered biomarkers remains the most challenging aspect of clinical proteomics, with the clinical laboratory being a preponderant figure to carry out the validation processes. The clinical laboratory must be critical at the moment of their validation, and for this they must undertake studies to evaluate the diagnostic sensitivity and specificity as well as the reproducibility of the biomarker in question; the design of the study and the analysis of data are crucial [27,28].

On the other hand, quality control systems are necessary procedures to validate the quality of the results. The majority of the problems found in proteomics are at the level of the preparation and manipulation of the samples, the fraction of these, and the analysis and evaluation of data; therefore, all these processes require supervision and the adequate use of the flows of work that allows to guarantee the veracity of the obtained results. Clinical laboratories that work with proteomic techniques should introduce as part of the routine the quality control in the different stages of the workflows used in the analysis of a sample. The quality control efforts are currently specific for the different laboratories, which leads to very different strategies to control the operation of the instrumentation and the analysis steps. An important challenge for clinical laboratories is to determine strategies that standardize the procedures to be followed in the analysis of the proteome and sensitize the scientific community to the importance of quality control to obtain results that are increasingly valid.

The Human Proteome Organization has produced protein samples available as a benchmark quality control test that can be applied to all workflows used in proteomics. In this way, information is provided that allows evaluating the general state of the instruments and the workflow used.

Clinical laboratories, associated with clinical services, and incorporating basic scientists, form the basis for improving the research infrastructure, so necessary, for the multidisciplinary research teams in biomedical sciences to function effectively with the fundamental objective of achieving a quality research.

The clinical laboratory should be aware of the rapid advances in the field of genomics and proteomics, and new generation technologies ought to be put at the service of diagnosis, because only then will it be in a position to take advantage to maximize the opportunities that this scientific revolution offers us.

1.5 CLINICAL APPLICATIONS OF PROTEOMICS

From a practical point of view, there are many applications that can be glimpsed in various clinical areas. The identification of the proteins that intervene in the different stages of a disease will help to understand the molecular bases and the nature of this anomaly; at the same time, these identified proteins can be used as biomarkers to establish the diagnosis or prognosis of the disease. Understanding the molecular processes of pathologies, such as cancer or autoimmune diseases, will contribute to institute more effective health policies that have an impact on the well-being of the population. It is also possible to identify new therapeutic targets for a better drug design and the monitoring of the effects of a substance in the treatment of a patient. Knowledge of the proteomes of different pathogens in interaction with their host will make possible a better understanding of the biology of the system and provide better tools for the control of the infectious diseases. Finally, MS is a technique that has now been used to conduct population genotyping studies and detect multiple pathogens simultaneously and at low cost.

The following discusses some of the areas in which proteomics has had a great impact:

1.5.1 Microbiology

In less than 10 years, the MALDI-TOF MS has gone from being a technology outside the diagnostic procedures of clinical microbiology to being a fully integrated procedure in daily clinical activity.

Bacterial identification is fundamental in the proper management of infected patients, especially those with severe infections hospitalized in critical patient units. The use of MALDI-TOF MS in microbiological identification in clinical laboratories has undoubtedly been one of the most important changes after the introduction of molecular methods. Its application in the bacterial and yeast identification in a clinical laboratory allows to shorten the response significantly, improve the accuracy of the identification in comparison with that of conventional methods, and reduce the high costs, regarding reagents, supplies, and handwork. At the same time, it has determined an important economic saving in health institutions by saving empirical antibiotic therapies [29,30].

Since its introduction in clinical microbiology laboratories, numerous studies have shown in recent years the advantages of MALDI-TOF MS for bacterial identification, demonstrating its usefulness in the identification of microorganisms from the colonies, as well as directly from positive blood cultures and urine samples [31,32].

In fact, the limitations of the MALDI-TOF MS mostly lie in the libraries of the reference databases, rather than the ability of the method to obtain reliable profiles of almost any bacterial microorganism. The same goes for yeasts; in the case of filamentous fungi the results are not as favorable, but again this seems to be a limitation more associated to the heterogeneity of the protein profiles generated by a filamentous fungus throughout its development and to the incapacity of databases, so far, to collect them.

It is worth highlighting the studies of the total proteome of the pathogenic microorganisms. These are very significant because it obtains information of the protein expression in a determined condition. It is also possible to obtain information of certain subproteomes, such as the proteins exposed in the membrane or to know which proteins trigger the immune response in the host. Particularly important is the membrane subproteome, which constitutes a set of proteins related to the infection. There are techniques for the detection of proteins exposed on the cell surface. The secreted proteins also constitute an important subproteome and can trigger the lysis of the host cells by means of toxins, which can influence the adhesion, colonization, and invasion, as well as the deviation of the host-immune response. The proteins of the outer membrane vesicles constitute another very important subproteome in the pathogens, as they show a powerful immunogenic reactions when released by the microorganism to the extracellular medium [33–35].

The possibility of using the MALDI-TOF MS to characterize pathogenicity factors, mainly toxins, has also been proposed. In the usual methods up to now, the protein peaks obtained are not characterized individually, because it is not necessary for the identification of microorganisms. However, this possibility exists and in fact, MALDI-TOF MS has been used for the detection of some bacterial toxins (*Clostridium perfringens* and *Staphylococcus aureus*).

The implementation of MALDI-TOF MS has revolutionized the area of clinical microbiology, clearly marking a before and an after since its incorporation into clinical practice.

1.5.2 Biomarkers

Biomarkers are biomolecules that are determined in a tissue or body fluid of a patient to define a disease at a molecular level and to follow or predict the response to a therapy for a certain disease. The ideal protein markers for clinical routine are those that can be detected in the serum, because the serum contains information about the proteome, which potentially reflects what is occurring in each tissue of the body.

Proteomics has been used in research studies in different areas of medicine. These studies can be classified according to the type of sample used, the pathology in question, or the type of pathologies that address.

In this way, studies of protein biomarkers have been developed from biological fluids, cell lines, or solid tissues, with the aim of establishing the diagnosis, follow-up, treatment, and prognosis of tumoral, autoimmune, and infectious pathologies, among others [36–38].

Fluid proteomics has been of great importance in the field of clinical research, given the ease of obtaining and processing samples, particularly serum, plasma, or whole blood [39]. The application of proteomic techniques to these samples has been used for the study of neoplastic, endocrinological, toxicological, rheumatic, neuromuscular diseases, among others [39−42]. Another example of fluid proteomics, although the sample is more difficult to obtain, is the analysis of the cerebrospinal fluid proteome. This has recently been used in the study of the etiopathogenic bases and the identification of biomarkers of neurological diseases, such as neuropsychiatric disorders, brain tumors, and lumbar pain syndromes [43,44].

In research with cell lines, proteomic techniques have been used to characterize the protein expression profiles in different types of these lines as a basis for future comparative experiments. Cell lines are variants derived from cancer cells, and unlike most cells isolated directly from organisms, these cells are able to divide indefinitely in cultures. Many studies have shown that the number of proteins identified in each proteome varies significantly, which is probably a reflection of the existence of differences in the range of expression of the different proteins in each cell line [36].

Human tumor cell lines are new models for the study of cancer; studies comparing the proteomes of different human tumor lines with their corresponding normal cell lines are used to identify disease markers or biomarkers that allow early detection, classification, and establishment of the prognosis of tumors, as well as to establish therapeutic targets to improve their treatment [40]. There are also studies aimed at evaluating the malignant potential of a tumor according to whether it expresses a specific protein, which allows us to define the best therapeutic strategy based on it.

Proteomics has also been used for the design, indication, and prediction of the response to drugs; it is what we know as pharmacoproteomics [45].

On the other hand, cell lines have been used classically to know the adaptive mechanisms to different types of stress. The development of proteomics has found a clear application in these studies, because these cell lines allow the identification of new proteins not related up to now with the establishment of a phenotype resistant to the conditions that cause cell stress.

Proteomic techniques have been applied more recently in the analysis of solid and fluid tissues for the studies of physiological or pathological situations (cancer, autoimmune diseases, infections, etc.). These studies with tissues or fluids have been used with similar objectives to studies with cell lines such as identification of biomarkers, search for therapeutic targets, and pharmacoproteomics.

Although the use of solid tissue samples in proteomics studies has the disadvantage of the inaccessibility of the sample, the information obtained in these can be determinant and sometimes the only or most accurate alternative [46]. Such may be the case of the study of selective mutations that alter the proteome of a tissue locally, which are not detectable in other compartments of the organism.

1.5.3 Immunoproteomics

Immunoproteomics is a potentially useful tool in the field of biomarker search for infectious diseases, especially in situations where the pathogen is difficult to grow in vitro. SELDI technology has been applied very recently in the identification of biomarkers to discern strains of *Helicobacter pylori* associated with different stages of the disease. In this way, "signatures" of serum antibodies are identified [47].

The identification and characterization of antigenic proteins is a requirement for the development of vaccines, because they have a great interest as a therapeutic alternative to treatment. Immunoproteomics is useful to provide general diagnostic proposals or for the selection of biomarkers.

1.5.4 Vaccines and drugs

The selection of targets for drugs and/or vaccines is essential for obtaining an optimal result in the control and/or treatment of many diseases. The data obtained from the proteome during infection can be exploited as a basis for the development of new antimicrobial chemotherapies and vaccines [48,49].

1.6 BIOINFORMATICS' APPLICATION TO PROTEOMICS

As its name suggests, bioinformatics is an area of research positioned at the interface between two sciences: computer science and biology. Bioinformatics arises in response to the need for computer tools that would allow managing, interpreting, and distributing the large volume of information obtained in the genome projects. Bioinformatics is also one of the techniques applied to the study of proteomics, through a set of tools used to interpret the data obtained with different MS methodologies.

Bioinformatics tools allow to identify a protein in a database from its peptide fingerprint or its spectrum, characterize all types of proteins, characterize protein variability (types of proteins, amounts, activity levels) in various cell types, etc. Bioinformatics are responsible for developing formulas and mathematical models that assist researchers in the comparison of sequences, conformational study of biomolecules, and resolution of phylogenies, among others. To carry out its objectives, bioinformatics has a series of databases and different software packages.

The development of new computer algorithms capable of intelligently interpreting the information generated by the MS will allow in the near future to apply proteomics to numerous clinical studies.

1.7 CHALLENGES OF PROTEOMICS

Proteomics is a challenge in itself, but from the methodological point of view, it is important to highlight the enormous potential in the permanent development of new tools applicable to prevention, diagnosis, treatment, follow-up, and prognosis in human pathology.

Getting to quantify each protein of interest from rapid MS methods is also a current challenge. This would gain in precision when establishing differential patterns. On the other hand, the design of "arrays" and "microarrays" of proteins and antibodies for their use on a large scale is in its beginnings; its use will allow facing the study of complex protein mixtures.

In the area of microbiology, once established its use for microbial identification, achieving through proteomic techniques, sensitivity profiles similar to conventional phenotypic profiles, with a speed and reliability close to those achieved in the identification is an important challenge, given the variety of resistance mechanisms and the diversity of proteins involved. A field still in the early stages of its exploration is the potential utility of MALDI-TOF electron multiplier in the typing of microorganisms, remaining yet an important standardization work, so that the MS can be an alternative tool to the molecular epidemiology methods.

1.8 CONCLUSIONS

The proteomic studies are acquiring in recent years a great relevance, fundamentally in what refers to its application to human pathology.

To this end, a large number of studies are being carried out on human plasma, tissues, and various biological fluids. The practical utility of the results obtained with proteomics, in relation to health, is very important. The discovery of protein markers of conditions such as cardiovascular, neurological, oncological, and metabolic, among others, has an immediate clinical application in the diagnosis, follow-up, and treatment of these diseases.

In the same way, proteomics is a truly useful tool in the study of infectious diseases, and proteomic analyses allow the realization of a basic study of the disease, facilitating the search for markers of infection or virulence. The results obtained have a translational applicability aimed at the diagnosis and prognosis of the disease, additionally allowing the development of new antimicrobial therapies and the progress in vaccination. The proteomic techniques not only allow the identification of new markers, but they can also be implanted as quick diagnostic techniques, being able to direct the treatment with greater speed.

REFERENCES

[1] Righetti PG, Castagna A, Antonucci F, Piubelli C, Cecconi D, Campostrini N, et al. Proteome analysis in the clinical chemistry laboratory: myth or reality? Clin Chim Acta 2005;357:123–39.
[2] Proteome Research. New frontiers in functional genomics. Berlin, New York: Springer; 1997.
[3] Plebani M. Proteomics: the next revolution in laboratory medicine. Clin Chim Acta 2005;357:113–22.
[4] Graves PR, Haystead TA J. Molecular biologist's guide to proteomics. Microbiol Mol Biol Rev 2002;66:39–63.
[5] Master SR. Diagnostic proteomics: back to basics? Clin Chem 2005;51:1333–4.
[6] Legrain P, Aebersold R, Archakov A, Bairoch A, Bala K, Berretta L, et al. The human proteome project: current state and future direction. Mol Cell Proteomics 2011;10(7) M111.009993.
[7] Gil C. The proteomic methodology, a tool for the function search. Actualidad SEM 2003;35:12–20.
[8] Li J, Kelm KB, Tezak Z. Regulatory perspective on translating proteomics biomarkers to clinical diagnostics. J Proteomics 2011;74(12):2682–90.
[9] Issaq HJ, Conrads TP, Janini GM, Veenstra TD. Methods for fractionation, separation and profiling of proteins and peptides. Electrophoresis 2002;23(17):3048–61.

[10] Gorg A, Weiss W, Dunn MJ. Current two-dimensional electrophoresis technology for proteomics. Proteomics 2004;4:3665−85.

[11] Aebersold R, Mann M. Mass spectrometry-based proteomics. Nature 2003;422 (6928):198−207.

[12] Mann M, Hendrickson RC, Pandey A. Analysis of proteins and proteomes by mass spectrometry. Annu Rev Biochem 2001;70:437−73.

[13] Karas M, Bachman D, Bahr U, Hillenkamp F. Matrix-assisted ultraviolet laser desorption of non-volatile compounds. Int J Mass Spectrom Ion Process 1987;78 (0):53−68.

[14] Schwamborn K, Caprioli RM. Molecular imaging of proteins using MALDI-TOF MS. Ann Chem 1997;69(23):3626−32.

[15] Issaq HJ, Conrads TP, Prieto DA, Tirumalai R, Veenstra TD. SELDI-TOF MS for diagnostic proteomics. Anal Chem 2003;75:148−55.

[16] Fenn JB, Mann M, Meng CK, Wong SF, Whitehouse CM. Electrospray ionization for mass spectrometryof large biomolecules. Science 1989;246(4926):64−71.

[17] Michalski A, Damoc E, Hauschild JP, Lange O, Wieghaus A, Makarov A, et al. Mass spectrometry-based proteomics using Q Exactive, a high-performance benchtop quadrupole Orbitrap mass spectrometer. Mol Cell Proteomics 2011;10(9) M111.011015.

[18] Eng JK, Mc Cormack AL, Yates LII JR. An approach to correlate tandem mass spectral data of peptides with amino acid sequences in a protein database. J Am Soc Mass Spectrom 1994;5(11):976−89.

[19] Hu Q, Noll RJ, Li H, Makarov A, Hardam M, Graham Cooks R. The Orbitrap: a new mass spectrometer. J Mass Spectom 2005;40(4):430−43.

[20] Andrews GL, Simons BL, Young JB, Hawkridge AM, Muddiman DC. Performance characteristics of a new hybrid quadrupole time-of-flight tandem mass spectrometer. Anal Chem Soc 1998;120(13):5442−6.

[21] Old WM, Meyer-Arendt K, Aveline-Wolf L, Pierce KG, Mendoza A, Sevinsky JR, et al. Comparison of lavel-free methods for quantifying human proteins by shotgun proteomics. Mol Cell Proteomics 2005;4(10):1487−502.

[22] MacBeath G. Protein microarrays and proteomics. Nat Genet 2002;32:526−32.

[23] Wu W, Slastad H, de la Rosa Carrillo D, Frey T, Tjonntord G, Boretti E, et al. Antibody array analysis with label-based detection and resolution of protein size. Mol Cell Proteomics 2009;8(2):245−57.

[24] Droit A, Poirier GG, Hunter JM. Experimental and bioinformatics approaches for interrogating protein-protein interactions to determine protein function. J Mol Endocrinol 2005;34:263−80.

[25] MacLean B, Tomazela DM, Shulman N, Chambers M, Finney GL, Frewen B, et al. Skyline: an open source document editor for creating an analyzing targeted proteomics experiments. Bioinformatics 2010;26(7):966−8.

[26] Perkins DN, Pappin Dj, Creasy DM, Cottrell JS. Probability-base protein identification by searching sequence database using mass spectrometry data. Electrophoresis 1999;20(18):3551−67.

[27] Ptolemy AS, Rifai N. What is a biomarker? Research investments and lack of clinical integration necessitate a review of biomarker terminology and validation schema. Scan J Clin Lab Invest Suppl 2010;242:6−14.

[28] Andre F, McShane LM, Michiels S, Ransohoff DF, Altman DG, Reis-Filho JS, et al. Biomarker studies: a call for a comprehensive biomarker study registry. Nat Rev Clin Oncol 2011;8(3):171−6.

[29] Seng P, Drancourt M, Gouriet F, La Scola B, Fournier PE, Rolain JM, et al. Ongoing revolution in bacteriology: routine identification of bacteria by matrix-assisted laser desorption ionization time-of-flight mass spectrometry. Clin Infect Dis 2009;49:543−51.

[30] Sauer S, Kliem M. Mass spectrometry tools for the classification and identification of bacteria. Nat Rev Microbiol 2010;8(1):74—82.

[31] Lotz A, Ferroni A, Beretti JL, Dauphin B, Carbonnelle E, Guet-Revillet H, et al. Rapid identification of mycobacterial whole cell in solid and liquid culture media by matrix-assisted laser desorption ionization time-of-flight mass spectrometry. J Clin Microbiol 2010;48:4481—6.

[32] Drancourt M. Detection of microorganisms in blood specimens using matrix-assisted laser desorption ionization time-of-flight mass spectrometry: a review. Clin Microbiol Infect 2010;16:1620—5.

[33] Rodríguez-Ortega MJ, Norais N, Bensi G, Liberatori S, Capo S, Mora M, et al. Characterization and identification of vaccine candidate proteins through analysis of the group A Streptococcus surface proteome. Nat Biotechnol 2006;24:191—7.

[34] Sinha S, Arora S, Kosalai K, Namane A, Pym AS, Cole ST. Proteome analysis of the plasma membrane of *Mycobacterium tuberculosis*. Comp Funct Genomics 2002;3:470—83.

[35] Dumrese C, Slomianka L, Ziegler U, Choi SS, Kalia A, Fulurija A, et al. The secreted Helicobacter cysteine-rich protein A causes adherence of human monocytes and differentiation into a macrophage-like phenotype. FEBS Lett 2009;583 1637—43.

[36] Moore GE, Minowada J. Historical progress and the future of human cell culture research. Hum Cell 1992;5:313—33.

[37] Le Naour F. Contribution of proteomics to tumor immunology. Proteomics 2001;1:1295—302.

[38] Yamamamoto T. The 4th Human Kidney and Urine Proteome Project (HKUPP) workshop. 26 September 2009, Toronto. Canada Proteomics 2010;10(11):2069—70.

[39] Teng PN, Bateman NW, Hood BL, Conrads TP. Advances in proximal fluids proteomics for disease biomarker discovery. J Proteome Res 2010;9(12):6091—100.

[40] Lage H. Proteomics in cancer cell research: an analysis of therapy resistance. Pathol Res Pract 2004;200:105—17.

[41] Boylan KL, Geschwind K, Koopmeiners JS, Geller MA, Starr TK, Skubitz PN. A multiplex platform for the identification of ovarian cancer biomarkers. Clin Proteomics 2017;14(34):1—21.

[42] Liotta LA, Petricoin EF. Serum peptidome for cancer detection: spinning biologic trash into diagnostic gold. J Clin Invest 2006;116:26—30.

[43] Zheng PP, Luider TM, Pieters R, Avezaat CJ, Van den Bent MJ, Sillevis Smitt PA, et al. Identification of tumor-related proteins by proteomic analysis of cerebrospinal fluid from patients with primary brain tumors. J Neuropathol Exp Neurol 2003;62:855—62.

[44] Wulvuhle JD, Sgroi DC, Krutzsch H, McLean K, McGarvey K, Knowlton M, et al. Proteomics of human breast ductal carcinoma in situ. Cancer Res 2002;62:6740—9.

[45] Jain KK. Role of pharmacoproteomics in the development of personalized medicine. Pharmacogenomics 2004;5:331—6.

[46] Chaurand P, Sanders ME, Jensen RA, Caprioli RM. Proteomics in diagnostic pathology: profiling and imaging proteins directly in tissue sections. Am J Pathol 2004;165:1057—68.

[47] Bernarde C, Khoder G, Lehours P, Burucoa C, Fauchère J-L, Delchier J-C, et al. Proteomic Helicobacter pylori biomarkers discriminative of low-grade gastric MALT lymphoma and duodenal ulcer. Proteomics Clin Appl 2009;3:672—81.

[48] Bambini S, Rappuoli R. The use of genomics in microbial vaccine development. Drug Discov Today 2009;14:252—60.

[49] Poland GA, Ovsyannikova IG, Jacobson RM. Application of pharmacogenomics to vaccines. Pharmacogenomics 2009;10:837—52.

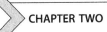

CHAPTER TWO

Basis of Mass Spectrometry: Technical Variants

Zdeněk Perutka and Marek Šebela

Department of Protein Biochemistry and Proteomics, Centre of the Region Haná for Biotechnological and Agricultural Research, Faculty of Science, Palacký University, Olomouc, Czech Republic

2.1 OVERVIEW

Chapter 2, Basis of Mass Spectrometry: Technical Variants, provides a definition of mass spectrometry (MS) and describes the history of this analytical technique from the very beginning. Then it summarizes basic components of a mass spectrometer, from which namely mass analyzers are further described in detail together with their combinations in hybrid mass spectrometers. Such instruments are designed to perform tandem MS, which has been used in chemistry and biology to identify and quantify compounds. All possible modes of tandem MS are mentioned as well as common ion activation techniques for the fragmentation of gas–phase ions, which is essential and occurs between different stages of mass analysis. Next, separation techniques hyphenated with MS are discussed, namely gas and liquid chromatography (GC and LC, respectively) and capillary electrophoresis (CE). Hyphenated systems are nowadays almost indispensable for investigations of complex biological samples containing proteins, lipids, or metabolites. Mass spectrometry imaging (MSI) is presented as a tool for direct analyses of biological tissues. The chapter is ended by a part devoted to data representation and management with a focus on free and open source software and data formats. The text is accompanied by a list of the cited literature.

2.2 MS: DEFINITION AND BASIC PRINCIPLES

MS is an analytical technique that measures masses of atoms and molecules after their conversion to charged ions by an ionization process.

The Use of Mass Spectrometry Technology (MALDI-TOF) in Clinical Microbiology.
DOI: https://doi.org/10.1016/B978-0-12-814451-0.00002-2

It is applicable to volatile or nonvolatile samples, polar or nonpolar, solid, liquid, or gaseous. Under specific conditions, MS measurements can also reveal structural or quantitative information. With a modern instrumentation of the current period, it is highly sensitive and typically provides a high resolving power and mass accuracy. Mass spectrometric data for the analyzed compounds (ions) are commonly expressed as the mass-to-charge ratio, m/z (the abbreviation should always be typed in italics), which is the mass of the ion divided by the number of its charges [1]. The total charge is $q = ze$, where e is the elementary charge that equals 1.602×10^{-19} coulomb (C). Hence the SI unit for the expression m/q is $kg \cdot C^{-1}$. In MS practice, however, the dimensionless m/z is used as it is numerically better related to the unified atomic mass unit (u) or dalton (Da). One unified atomic mass unit (1 u) or 1 Da is defined as $1/12$ of the mass of a single ^{12}C atom. The Thomson unit (Th) commemorating Sir Joseph J. Thomson [2] has been used only rarely for m/z values; it is now deprecated. The terms mass spectrometer, mass spectrograph, and mass spectroscope should not be confused with each other. They are not synonymous and the latter two represent rather a historical reminiscence [1]. A mass spectroscope differed from a mass specrograph in its construction as the beam of ions was directed onto a phosphor screen and not to a photographic plate (see in the text section on the history of MS, Section 2.3). The term mass spectroscopy is now discouraged due to the possible confusion with light spectroscopy.

Due to its essential properties, MS plays an important role in many fields of human activities including science, medicine, industry, and public safety sector [3]. In physics, it allows to measure data on particles, elements, materials, energy transfers, dynamics of various processes, etc. It is advantageous for evaluating results of theoretical physical predictions and calculations. MS is probably the most versatile analytical technique in chemistry including biochemistry. Chemists profit, for example, from the accurate measurements of atoms and molecules, structural and quantitative analysis of synthetic and natural compounds, drugs, toxins and pollutants, investigating chemicals processes or fundamentals of gas-phase ion chemistry. Chemical compounds, isotopes, as well as isotope ratios may attract interests of geologists, archeologists, or space researchers. MS is always at hand as an analytical tool. For complex real samples, MS is usually coupled with high-resolution separation techniques (e.g., GC, LC or CE). Medicine and life sciences utilize such hyphenated techniques for a simultaneous separation and detection of compounds in mixtures originating from biological tissues and fluids. MS is applicable for an efficient amino acid sequencing and related identification of peptides and proteins, quantification of proteins, determination of posttranslational protein modifications, analyses of protein—protein interactions, and composition of protein complexes [4]. Also short oligonucleotides can be efficiently sequenced using MS. Biomolecules (e.g., proteins, peptides, lipids, oligosaccharides, and various metabolites) are amenable to MS-based structural characterizations. For proteins as large molecules, there are indirect methods available involving a covalent crosslinking of adjacent parts of the meandering polypeptide chain or deuterium exchange experiments, which uncover solvent-accessible regions of the three-dimensional structure. These structural studies are helpful to discover functional aspect of proteins. Based on acquiring peptide/protein profile mass spectra using intact cells or spores, MS allows identifying microorganisms, which is useful especially for pathogenic species in clinical microbiology. Industrial applications of MS commonly serve for monitoring of process streams, e.g., in the petroleum or pharmaceutical industry. This introductory overview of the applicability of MS is definitely not complete. We may expect that with future developments, MS will expand to yet unexplored territories to offer new perspectives on solving analytical problems.

2.3 A BRIEF EXCURSION TO THE HISTORY OF MS AND RELATED NOBEL PRIZES

For over than 100 years, MS has gradually been introduced into many scientific disciplines as a reliable tool for molecular analysis. In the light of its current wide applicability, it seems interesting that the history of this analytical technique began in the late 19th century in connection with electrical discharges and electrode rays. Eugen Goldstein (1850—1930), a German physicist, was one of early investigators of discharges in rarefied gases and is credited with his contribution to the discovery of proton [5,6]. He was a student of the famous physician and physicist Hermann Ludwig Ferdinand von Helmholtz (1821—94). In 1886, he reported rays in gas discharges under a low pressure, which traveled from the anode to a perforated cathode through channels in the perforation, exactly opposite to the direction of the negatively charged cathode rays (i.e. electrons) discovered and studied previously by other scientists (Johann Wilhelm Hittorf, Julius Plücker). Goldstein introduced the term "Kanalstrahlen" (canal rays) for the positively charged anode rays. Wilhelm Wien (1864—1928), another student of Helmholtz, discovered that these rays required much stronger magnetic fields to be deflected compared to cathode rays. He constructed a device that separated particles constituting the positive rays according to their charge and mass [3,7]. Today, most people in the field regard Sir Joseph J. Thomson (1856—1940), the discoverer of electron and Cavendish Professor at the University of Cambridge, as the father of MS. He began with studies on canal rays around 1906 when he was awarded the Nobel Prize in Physics for investigations on the conduction of electricity by gases. He noticed that positive rays showed varying masses reflecting the presence of different gases in the discharge. Thomson is credited with his realization that this new technique would play a profound role in the field of chemical analysis [3,8,9]. In the next two decades, the development of MS occurred in connection with the names Aston, Dempster, and Nier [3]. Francis William Aston (1877—1945) was an assistant of Thompson and in 1912 observed the existence of two isotopes of neon (^{20}Ne and ^{22}Ne) [8,9]. This was the first evidence of isotopes of a stable element. Early devices that measured positive ions were called mass spectrographs as they recorded spectra on a photographic plate. Aston constructed more spectrographs with a gradually increasing performance. They comprised two sectors to improve the resolving power: the electric sector provided energy selection and the other one, magnetic sector, allowed mass separation [10]. With the spectrographs, Aston was able to study isotopes of many elements and was awarded the Nobel Prize in Chemistry in 1922. Arthur Jeffrey Dempster (1886—1950), a Canadian—American physicist at the University of Chicago, built the first modern mass spectrometer in 1918 producing ions (homogeneous in energy) also from solid materials [11]. In the 1930s, Dempster built instruments containing both a direction and velocity focusing. His experimental work led finally to the discovery of the uranium isotope ^{235}U [12].

During the 1940s, MS played an important role in the Manhattan Project, a wartime program involving a preparative-scale separation of the fissionable ^{235}U isotope for atomic bombs [3]. Sector mass spectrometers (calutrons) developed by Ernest Orlando Lawrence (1901—58) at the University of California in Berkeley were used in the uranium enrichment plant in Oak Ridge, Tennessee. Alfred Otto Carl Nier (1911—40) was another American physicist participating in the Manhattan project, yet he contributed considerably to the conversion of MS from a device used primarily for investigating chemical elements and their isotopes to an analytical tool for chemists. He constructed a simple mass spectrometer with a 60-degree-angle magnetic sector. Interestingly, this instrument was soon applied to track metabolic processes in bacteria via carbon dioxide [13]. MS became ultimately recognized by chemists. The first commercial mass spectrometers for the petroleum industry (to analyze the refining process) were available through the Consolidated Engineering Corporation in 1943 [3,14]. MS applications in organic chemistry commenced in the 1950s and became increasingly popular in the 1960s and 1970s [3]. With its continuous development, new mass analyzers were introduced in this period together with new ionization techniques, e.g., chemical ionization (CI). Wolfgang Paul (1913−93), a physicist from Bonn in Germany, introduced the concept of a quadrupole mass analyzer and ion trap to separate ions without a magnetic field [15]. In 1989, he was awarded for this achievement by sharing the Nobel Prize in Physics with Hans Georg Dehmelt, another researcher on ion traps [3].

MS has emerged as a standard tool for investigating organic compounds [4]. In the 1960s, tandem MS was developed [16], which allowed structural analyses based on reading information from fragmentation patterns of dissociated precursor ion (see further in this chapter). The analyses of compounds in complex mixtures have become much more feasible since the introduction of hyphenated techniques: gas chromatography (GC)−MS in the 1960s and LC-MS in the 1970s. However, for a long time, the use of MS for biological samples with fragile and nonvolatile compounds was marginal because of the lack of suitable ionization techniques [3]. To produce ions from compounds such as porphyrins, oligosaccharides, and peptides, fast atom bombardment (FAB) ionization was introduced in 1981 [17], which was outperformed by electrospray ionization (ESI) [18] and matrix-assisted laser desorption/ionization (MALDI) [19,20] at the end of the 1980s. The discovery of the soft ionization techniques (now typically applied to peptides, proteins, and various metabolites) had an enormous impact on the use of MS in biology and new life science disciplines such as proteomics [3]. This was reflected in awarding the Nobel Prize in Chemistry in 2002 to John Bennett Fenn and Koichi Tanaka. The last two decades provided many improvements in MS instrumentation and data systems. Desorption electrospray ionization (DESI) for sampling under ambient conditions [21] or Orbitrap mass analyzer for high-resolution measurements [22] are two examples to be mentioned at this place.

2.4 BASIC COMPONENTS OF A MASS SPECTROMETER

Fig. 2.1 shows a simplistic scheme of a mass spectrometer. Each instrument consists of several major components: (1) an ion source with a sample inlet, (2) a mass analyzer, (3) a detector, and (4) a data recording and processing system. Indispensable parts are also a vacuum system and controlling electronics [3]. The inlet transfers a sample into the ion source, where sample molecules are converted into gas-phase ions. The ionization chamber is kept in a vacuum, which allows the ions to travel to the analyzer by involving a magnetic or electric field force. There are many possibilities how to ionize chemical or biochemical compounds, some of them are introduced in the next part of this chapter. Based on the amount of the transferred energy, there are soft and hard ionization techniques producing ions with low or high internal energy, respectively [23]. Mass analyzer explores and sorts ions with different properties. It can be used for an overall analysis, or it may function as a filter to select only specific ions for the subsequent analysis and detection. Basic principles of mass analyzers include, e.g., magnetic or electric fields to manipulate the trajectory of ions, utilizing differences in their velocities (when measuring the time of passing the distance in a flight tube) or resonance frequencies (when

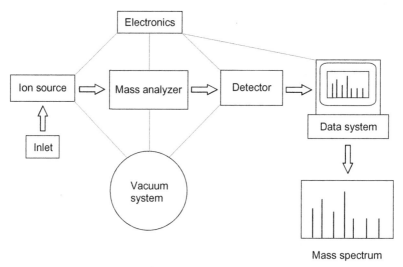

Figure 2.1 *Basic composition of a mass spectrometer.* *Adapted from Dass C. Basics of mass spectrometry. In: Fundamentals of contemporary mass spectrometry. Hoboken, NJ: John Wiley & Sons, Inc.; 2007. pp. 3–14.*

the ions are trapped inside a trap analyzer and then released) and the principle of ion mobility in the gas phase, which is proportional to the collisional cross section of the respective ion [24,25]. Combining different analyzers inline to separate or fragment and subsequently determine the m/z of the selected ions is a way for how to increase the effectivity and versatility of commercial instruments (see further in this chapter).

The function of a detector is to "visualize" ions, which is done by recording their abundance [25]. There is a recordable electric signal produced, which can be amplified, processed, and displayed by the data system. There are two detector categories: either they detect ions sequentially at one point (focal point detectors) or simultaneously along a plane (focal plane detectors) [25]. Today, common detectors are based on an electron multiplier (EM) [26], which is constructed as a series of discrete dynodes or a continuous dynode (i.e. channel electron multiplier, CEM, in which the surface represents an array of continuous electrodes). Multichannel plate detector is a multichannel version of CEM. The EM principle resides in a multistage electron-releasing cascade initiated by incident ions (coming from the analyzer). In Fourier-transform ion cyclotron resonance (FT-ICR) and Orbitrap mass spectrometers, the detected signal is produced as a record of the image current generated by the oscillating ions inside the analyzer [26]. The most important detector characteristics are its sensitivity, accuracy, resolution, response time, and stability. It is also desirable to have a wide dynamic range and low noise level. The analyzer and detector must be both kept in a high vacuum to work properly. The pressure inside the analyzer reaches around 10^{-8} mbar [3]. To obtain such a value, the spectrometer is equipped with two pumps, typically an auxiliary mechanical membrane or rotary pump, and a turbomolecular pump generating the high vacuum [27]. The analog signals from the detectors are converted into digital information by digitizers. Powerful computers with a large data storage capacity are required for modern MS instrumentation.

2.5 IONIZATION TECHNIQUES IN BIOLOGICAL MS

The classical ionization methods in MS include, for example, electron ionization, CI, photoionization, and FAB. These techniques are not widely used for studying biological materials but represent pioneering originals, from which many other ionization methods were adapted. The progress in the analysis of biological molecules (including biopolymers such as proteins) has largely been stimulated by the introduction of ESI and MALDI [18–20]. In both cases, the production of ions is achieved with a minimal fragmentation under gentle experimental conditions. A further progress has come with the development of ambient ionization techniques, which facilitate the analysis from real objects outside the mass spectrometer without any extraction of the molecules of interest or pre-treatment of the object itself. DESI and direct analysis in real time (DART) have been reported as the main commercially available representatives of this emerging group [28].

ESI is the most widely used ionization technique for the analysis of samples in a liquid form. In the respective ion source, a sample is ionized at the outlet of the capillary, on which a high voltage is applied (Fig. 2.2). Therefore, these devices can be coupled online to separation techniques in the liquid phase such as LC. The sprayed droplets are dried continuously, and their size is reduced by the stream of a nebulizing gas. The chromatographic solvent (salt-free) carries sample molecules, evaporates rapidly, and the molecules remain inside droplets, which gradually diminish in time. Once the repulsive forces between the molecules overcome the droplet surface tension, it bursts and the charged molecules are released to enter the analyzer [30]. ESI is characteristic by the production of multiply charged ions [18]. This is advantageous for resolving large molecules in mass-range limited analyzers (such as quadrupoles) as they provide lower m/z values. Atmospheric pressure chemical ionization (APCI) represents another soft ionization technique similar to ESI. It is suited to analyze relatively less polar, nonpolar, and heat-stable compounds with masses up to 1500 Da in LC effluents. The ion source contains a heated nebulizer probe, and the emerging droplets and nebulizer gas are converted to a gas stream containing analyte molecules. The ionization region contains a corona discharge. Positive ions of the analyte are formed by a proton transfer from water clusters [31]. The atmospheric pressure photoionization (APPI) is a variation of APCI, in which the initial ionization is achieved by photons [32].

In the case of MALDI (Fig. 2.2), the sample must be first overlaid or mixed with a matrix compound solution and placed onto a conductive target plate. The MALDI probe is introduced into the ion source and irradiated—typically with an ultraviolet (UV) laser. A part of the sample

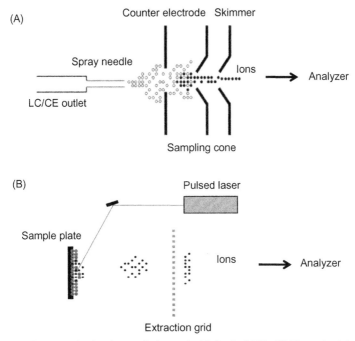

Figure 2.2 *Common ionization techniques in biological MS.* (A) The principle of ESI. (B) MALDI. *Adapted from Aebersold R, Mann M. Mass spectrometry-based proteomics. Nature 2003;422:198—207 [29].*

evaporates when the laser energy is absorbed by the matrix. Then the excited matrix ionizes analyte molecules (by definition, analyte is a chemical constituent of a sample that is of interest in an analytical procedure) via a proton transfer. The reverse proton transfer is also possible [33]. This process usually generates singly charged ions. MALDI MS has successfully been applied to the analysis of biomolecules (e.g., proteins, peptides, oligonucleotides, lipids, and sugars), large organic molecules, and polymers [34]. Small organic molecules with a UV-absorbing chromophore such as α-cyano-4-hydroxycinnamic acid (for peptides), 2,5-dihydroxybenzoic acid (for peptides and carbohydrates), sinapinic acid (for proteins), or 3-hydroxypicolinic acid (for oligonucleotides) are routinely used as matrices.

More than 40 ambient ionization technique for MS have been reported, mostly based on the principles of solid—liquid extraction, plasma-based ionization, and non-laser or laser desorption [28]. DESI allows acquiring mass spectra with ordinary samples in their native environment. Small charged solvent droplets are sprayed and directed at the sample mounted in a holder. As the droplets collide with sample surface structures, ions are released and transferred into a conventional mass spectrometer [35]. DESI is well suited for a rapid tissue analysis [36]. In DART, a gas (typically helium) flows through an electrical discharge, which yields ionized gas, electrons, and excited state atoms/molecules (metastables). With helium, the dominant mechanism of the production of positive analyte ions involves the formation of ionized water clusters and then a proton transfer [37]. DART has been used to analyze a wide range of analytes with a low molecular mass including metabolites, hormones, drugs, synthetic organic compounds, and pigments. Common samples include body fluids or tissues, foods, and beverages.

2.6 MASS ANALYZERS

Mass analyzer is the heart of a mass spectrometer and largely influences the performance of the instrument. Each mass analyzer separates ions according to their m/z ratio and focuses the resolved ions to facilitate their detection. In this regard, the function of this component is similar to the monochromator and lens of a standard spectrophotometer [25]. Currently, several types of mass analyzers are available, which utilize different principles of distinguishing ions. They can be evaluated based on the following parameters (Table 2.1): mass range, resolving power (the ability to distinguish ions that differ only slightly in their mass), mass accuracy (the measured error compared to the accurate mass), sensitivity, speed (how many spectra are acquired in a time unit), linear dynamic range (the range over which ion signal is linear with analyte concentration), transmission efficiency, adaptability (with respect to outfitting with an ionization technique and coupling with a separation technique), and tandem MS capability.

Sector mass spectrometers have been used for the longest time [38]. In principle, they are either single focusing (Fig. 2.3) or double focusing and use a magnetic field or a magnetic plus electrostatic field, respectively. The magnetic analyzer separates ions of different m/z into different beams by bending their trajectories (this is a directional focusing); the electrostatic analyzer selects ions according to their kinetic energy (ions of the same energy are focused at a single point). The double focusing mass

Table 2.1 A comparison of mass analyzers

	Magnetic	Quadrupole	QIT	LIT	TOF	FT-ICR
Mass range (Da)	15,000	4000	4000	4000	Unlimited	$>10^4$
Resolving power	10^2–10^5	4000	10^3–10^4	10^2–10^5	15,000	$>10^6$
Mass accuracy (ppm)	1–5	100	50–100	50–100	5–50	1–5
Abundance sensitivity	10^6–10^9	10^4–10^6	10^3	10^3–10^5	up to 10^6	10^2–10^5
Speed (Hz)	0.1–20	1–20	1–30	1–300	10^1–10^6	10^{-2}–10^1
Efficiency (%)	<1	<1–95	<1–50	<1–99	1–100	<1–95
Dynamic range	10^9	10^7	10^2–10^5	10^2–10^5	10^2–10^6	10^2–10^5
MS/MS capability	Excellent	Great	Great	Excellent	Great	Great
LC(CE)-MS adaptability	Poor	Excellent	Excellent	Excellent	Good	Good

Source: Adapted from Dass C. Mass analysis and ion detection. In: Fundamentals of contemporary mass spectrometry. Hoboken, NJ: John Wiley & Sons, Inc.; 2007. pp. 67–117.

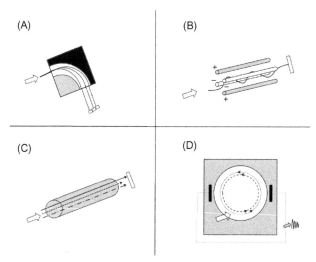

Figure 2.3 *Mass analyzers.* (A) Magnetic sector; (B) quadrupole; (C) linear flight tube (TOF); (D) FT-ICR. *Adapted freely from Dass C. Mass analysis and ion detection. In: fundamentals of contemporary mass spectrometry. Hoboken, NJ: John Wiley & Sons, Inc.; 2007. pp. 67–117.*

spectrometers are characterized by a high resolution and mass accuracy [25]. They are suitable, e.g., to study chemical reactions. On the other hand, they are not applicable for a coupling of the instrument with LC. The most common mass analyzer in mass spectrometers is probably a quadrupole [25]. It comprises four parallel metal rods representing electrodes (Fig. 2.3). There are direct current (DC) and radio frequency (RF) potentials applied to these electrodes, which create a high-frequency oscillating electric field. Under a certain combination of defined DC and RF potentials, ions of a specific m/z pass by a stable vibratory motion through the analyzer to reach the detector; other ions follow unstable trajectories and they are deflected. A mass spectrum is then obtained by changing the applied potentials when keeping their ratio constant. Standard quadrupoles provide a relatively low mass range, resolution, and accuracy (Table 2.1). Nevertheless, quadrupole-based instruments can be purchased at a reasonable price; they are sensitive and well suited for ESI and coupling with GC or LC, primarily for the analysis of low-mass compounds in chemistry and biology [39].

A time-of-flight (TOF) mass analyzer separates ions based on the difference in their velocities (Fig. 2.3). It is a long field-free flight tube and has traditionally been combined with a MALDI source [40]. Ions, produced in pulses, drift in the flight tube, and their velocities are in an inversed function to the square root of the respective m/z values. Classical linear TOF-MS has a poor resolution and is incompatible with continuous ion-beam sources such as electrospray (the latter can be overcome by the use of an orthogonal TOF). To minimize the spatial distribution and kinetic energy spread of ions, and increase resolution in consequence, both delayed extraction and reflectron (reflector) devices are added [25]. The ions formed in the ion source are accelerated by a delayed application of the electrical field to even out their different starting velocities. The reflectron is an energy corrector placed at the end of the flight tube (Fig. 2.4). It consists of ring electrodes with a gradually increased repelling potential, which make a barrier: the higher kinetic energy of an ion,

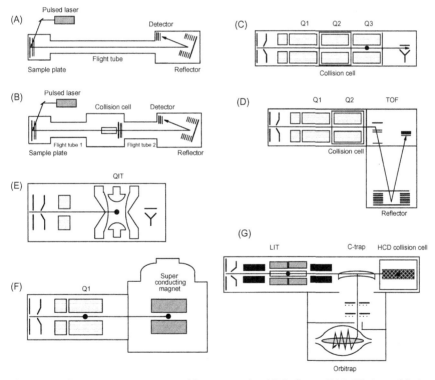

Figure 2.4 *Mass spectrometers used in proteomics.* (A) Reflector TOF; (B) time-of-flight reflector time-of-flight (TOF-TOF); (C) triple quadrupole (QqQ) or LIT; (D) quadrupole time-of-flight (Q-TOF); (E) QIT; (F) FT-ICR; (G) LIT-Orbitrap, HCD stands for higher energy collisional dissociation. *Adapted from Aebersold R, Mann M. Mass spectrometry-based proteomics. Nature 2003;422:198−207 [29], except for Panel (G), which is based on a picture by the vendor and made in the style of the others.*

the higher time spend in the reflectron. TOF analyzers theoretically have an unlimited mass range (but in practice it is restricted); they offer a high-spectrum acquisition rate and sensitivity (Table 2.1).

A Paul trap or quadrupole ion trap (QIT) consists of a central ring electrode and two end-cap electrodes (Fig. 2.4). There is an RF potential applied to the ring electrode when the end-cap electrodes are maintained at ground potential. Ions of a broad m/z range are trapped due to the oscillating three-dimensional electric field inside the analyzer, and their motion is then manipulated by increasing the amplitude of the RF potential [41]. This forces the trapped ions to become sequentially unstable, which is accompanied by ejecting them out of the trap for detection. QITs are simple to operate, relatively inexpensive, sensitive, and useful to conduct multistage MS experiments (MS^n) for structural studies. The major drawback is their poor mass accuracy and low dynamic range (Table 2.1). A linear ion trap (LIT) or two-dimensional QIT is made of four parallel rods with hyperbolic profiles; each rod is cut into three axial sections and there is a slit in one of the central rod sections to eject ions (Fig. 2.4). LIT can work as a selective filter or a real trap. Similar to QIT, mass analysis of the trapped ions is performed in the mass-selective instability mode by increasing the RF potential [42]. Compared to QIT, LIT has a higher trapping efficiency, ion storage capacity, and scan speed.

FT−ICR-MS was introduced in the 1970s [43]. The mass analyzer is formed by a cell (Penning trap) placed in a strong magnetic field (Fig. 2.3). The cell is composed of three pairs of opposite plates. Each pair has a different function (trapping, excitation, or detection). Ions are trapped in the cell and excited by an oscillating electric field orthogonal to the magnetic field. The ions excited due to a resonance-based energy transfer rotate in phase-coherent packets and are detected by measuring the image current at the detection plates. The resulting complex time-domain signal is processed to a frequency-domain representation by a Fourier transform. The masses of the ions are resolved by their ion cyclotron frequencies [43]. A typical attribute of FT-ICR-MS is a remarkably high resolving power and a superior mass accuracy. Orbitrap is a new mass analyzer, which was introduced in the 1990s, first commercial instruments then appeared in the 2000s [22]. It is an ion trap with a coaxial barrel-like outer electrode (it is split into halves by an insulating ring) and an axial spindle-like central electrode (Fig. 2.4). Contrary to a conventional ion trap mentioned above, there is no RF potential or magnetic field to keep ions inside an Orbitrap. Instead, ions are trapped in a pure electrostatic field and forced to cycle in elliptical trajectories around the central electrode; they also move forward and back along the electrode (z-axis). Together a spiral movement pattern is created. The axial component of these oscillations is related to the m/z values of the ions and can be detected as an image current induced between the axial halves of the outer electrode. FT is employed to get oscillation frequencies resulting in accurate m/z values. Another advantage is the high resolution available competing with that of FT-ICR instruments (Table 2.1) and largely surpassing orthogonal TOF analyzers [44].

2.7 TANDEM MS

Tandem MS has been used in chemistry and biology to identify and quantify compounds. It is based on multiple stages of mass analysis, which are coupled either in time or space [45]. If there are two stages, the abbreviation MS/MS or MS^2 is used. For more stages (feasible with ion-trapping instruments), MS^n is a general abbreviation. This technique has largely been perfected since its introduction in the late 1960s. Classical pioneering works included the development of collision-induced dissociation [16] and introduction of the triple quadrupole mass spectrometer [46]. In biological experiments, which utilize the soft ionization techniques ESI and MALDI, abundant structural information can be obtained only by MS/MS [47]. For example, MS/MS-based sequencing analyses of peptides in digests are necessary to get unambiguous data for protein identification. In research projects as well as routine analyses, often there is a need to select ions of a given mass (=precursor ions) for their activation, fragmentation, and mass analysis of the fragmentation products (=product ions). This is commonly applied for elucidating the chemical structure of unknown compounds, identification of target components of complex mixtures, and studying fragmentation pathways. Selective detection of ions providing a given fragment or losing a given neutral is also possible with an appropriate instrumentation [45,48]. Many instruments differing in their construction have become available for this purpose (Table 2.2). There are two types (Fig. 2.4): either with a tandem-in-space setup, where the individual stages of MS/MS are carried out in separate regions (mass analyzers), or a tandem-in-time arrangement, with all steps

Table 2.2 Common instruments applicable for tandem MS

	Q-Q-TOF	TOF-TOF	FT-ICR	Q-Q-Q	QQ-LIT
Resolving power	Good	High	Very high	Low	Low
Mass accuracy	Good	Good	Excellent	Medium	Medium
Dynamic range	Medium	Medium	Medium	High	High
ESI availability	Yes	–	Yes	Yes	Yes
MALDI availability	Optional	Yes	Optional	–	–
Identification	Good	Good	Excellent	Possible	Possible
Quantification	Excellent	Good	Good	Excellent	Excellent
Throughput	High	Very high	High	High	High
Detection of modifications	Possible	Possible	Possible	–	Excellent

Source: Adapted from Domon B, Aebersold R. Mass spectrometry and protein analysis. Science 2006;312:212–17 [49].

performed in the same analyzer by employing a temporal sequence of events [45]. Tandem-in-space instruments comprise several mass analyzers of the same type (e.g., multisector magnetic analyzers, triple quadrupole, TOF/TOF) or they have a hybrid design, in which different mass analyzers are coupled (e.g., magnetic sector–quadrupole, magnetic sector–TOF, quadrupole–TOF, quadrupole–LIT, and quadrupole–FT-ICR). The tandem-in-time variant is the case for ion-trapping instruments such as QIT, LIT, and FT-ICR [48].

There are four scan modes possible on tandem mass spectrometers: product ion scan, precursor ion scan, neutral loss scan, and selected reaction monitoring (SRM) [45]. All four scan modes are available with magnetic sector– and quadrupole-based instruments. TOF/TOF and ion-trapping devices are applicable for the product ion scan only, which is anyway the most common MS/MS experiment: here the selected ions are passed into the collision cell, activated, and induced to fragment [48]. The product ions are then analyzed and the fragmentation data are utilized to deduce structural information of the precursor ion. The precursor ion scan (also parent scan) is done in such an arrangement that the analyzer beyond the collision cell is set to pass exclusively those ions showing a particular (and selected) m/z value. Parent ions, which pass through the first analyzer (a quadrupole for instance), are detected only if they fragment in the collision cell to produce the selected product ion. The neutral loss scan, as in the case of the precursor ion scan, represents a setup where both analyzers are scanned together with a constant m/z difference. This allows recognizing all ions, which lose a given neutral fragment upon the fragmentation. The SRM approach is similar to acquiring a product-ion scan spectrum, and it is very useful for quantitative measurements because of its specificity. Instead of recording the complete spectrum of fragments, only a specific precursor—product pair is monitored (which means to detect a unique product ion) [50]. Monitoring more than one precursor-to-product transition is expressed by the term multiple reaction monitoring (MRM). Parallel reaction monitoring (PRM) takes the advantage of high-resolution MS. It is based on a quadrupole-Orbitrap platform. Unlike SRM/MRM, which provides one transition at a time, PRM performs parallel detection of all transitions (without the need to select a particular ion pair) in a single analysis. The Orbitrap analyzer scans all product ions with high resolution and high accuracy resulting in the elimination of the background interference and improvement of the detection limit and sensitivity [51].

Ion activation techniques are essential in MS/MS [52]. This is a brief overview: (1) the most common is collision-induced dissociation (CID) also named collisionally activated dissociation (CAD), for which precursor ions are excited by collisions (this can be performed in a high- or low-energy regime) with atoms of an inert gas such as helium or argon; (2) surface-induced dissociation (SID), available on a variety of instruments, where ion activation is achieved by collision with a solid surface (e.g., a metal plate); (3) absorption of UV/infrared photons or absorption of heat: UV photodissociation (PD), infrared multiphoton dissociation (IRMPD), or blackbody-induced radiative dissociation (BIRD), respectively; (4) electron-capture dissociation (ECD), which involves an excitation of protonated precursor by the capture of a low-energy electron and subsequent fragmentation of the resulting odd-electron ion; (5) electron-transfer dissociation (ETD)—the process is similar to ECD but uses an ion—ion reaction where anthracene anions are employed as electron donors. ECD has been primarily a part of FT-ICR instruments and it is solely applicable to the ESI-produced multiply protonated ions. Its applications include peptide sequencing, oligonucleotide sequencing, or identification of protein sites carrying posttranslational modifications such as phosphorylation, glycosylation, and others. Similarly, ETD is typically used to fragment multiply charged peptide ions for studying posttranslational modifications with ion traps.

2.8 SEPARATION TECHNIQUES HYPHENATED WITH MS

To analyze complex samples of the real world such as those from biological tissues, technological processes, and environmental control, the online coupling of MS with a separation device has long been a very successful approach. Commonly, GC, LC, or CE are used, and the respective coupling provides the possibility to increase the information gathered about sample components because of combining the resolving power of the separation technique and performance (particularly with respect to sensitivity and dynamic range) of a mass spectrometer [53]. The physical interconnection between the two instruments requires an interface, which delivers the separated sample components into the ion source. With a proper arrangement, both the resolution of the separation and performance of the mass spectrometer should remain virtually unaffected. The pressure mismatch (in GC–MS) and solvent incompatibility (in the case of LC and CE coupling) are the main issues to be dealt with. Not all mass spectrometers are suitable for the online coupling. Only those, which are characterized by a high scan speed and can tolerate high pressures (e.g., quadrupoles, ion traps, TOF analyzers) are considered optimal [54].

GC–MS is well suited for small in size, volatile, and thermally stable molecules. For larger diameter packed columns and higher carrier gas flows, there are interfaces available such as an open split interface, a jet separator, and a molecular effusion interface [55]. Capillary GC columns can be coupled directly to MS (because a low carrier gas flow is used) and provide a high-resolution separation of complex mixtures. LC and CE are suitable for the separation of mixtures with nonvolatile and thermally labile compounds. The introduction of ESI was an excellent opportunity to couple LC and also CE (as an approach complementary to LC) with MS. As a result, LC-ESI-MS in particular has become the method of choice for many applications including identification and quantification analyses of peptides, lipids, metabolites, drugs, and pesticides [56]. All sizes, from standard analytical high-performance liquid chromatography columns to capillary and nanoscale columns, are applicable for coupling with ESI—MS as there are different types of ESI sources and interfaces available for a wide range of solvent flow rates. Generally, smaller diameter columns are advantageous as they provide increased efficiency of separation. For many biochemical and biological studies, where sample amounts are often limited, capillary/nanoscale columns are applied [54]. Nanoscale columns are commonly operated at the flow range of $50-500$ nL $\cdot$ min^{-1} and attached directly to a nanospray ion source. With analytical and other larger diameter columns operated at higher flow rates (at the level of 100 µL to 1.0 mL $\cdot$ min^{-1}), a conventional ESI source (operates at a flow rate of $1-10$ µL $\cdot$ min^{-1}) can be used after a postcolumn flow splitting (e.g., 100:1). LC has also been coupled with MALDI MS [57]. Several constructs have been tried to make online combinations with a postcolumn mixing of the eluate with matrix solution (e.g., using a rotating wheel interface). Nevertheless, an offline approach with depositing eluate fractions on MALDI target in a spotter device after their mixing with a matrix solution has been successfully commercialized [56,57].

In proteomics, for example, reversed-phase LC (RP-LC) on a C18 (octadecyl hydrocarbon-bonded silica stationary phase) column is a standard technique to separate peptides prior to their MS/MS sequencing [54,58]. Besides this mode of separation, affinity chromatography (utilizing a specific affinity interaction of a biomolecule with a ligand immobilized in the stationary phase) or supercritical fluid chromatography (in which the mobile phase is a gas, such as carbon dioxide, liquid-like at a pressure above its critical pressure) can be coupled with MS. When the complexity of a sample is too high, one-dimensional LC is not sufficient to resolve optimally its components [54]. Then a two-dimensional chromatography is chosen, which combines two orthogonal separation steps (the meaning is that these steps represent completely different principles of separation). The second dimension must be faster and MS-compatible solvents have to be used to allow ionization. The most common approach is based on ion-exchange LC and RP-LC, but also size-exclusion chromatography, affinity chromatography, or chromatofocusing can be combined with RP-LC in the second dimension.

Finally, CE has been used in coupling with MS as an alternative to LC-based separations. Currently it is performed in different formats, including capillary zone electrophoresis (the most popular separation technique of this type), capillary gel electrophoresis, capillary isoelectric focusing, capillary isotachophoresis, micellar electrokinetic chromatography, and capillary electrochromatography [59]. The separation is frequently interfaced with ESI or APCI, and the three most practical designs are a sheathless interface, sheath–flow interface, and liquid–junction interface [60]. The CE effluent can also be deposited on MALDI target plate for an offline MS analysis.

2.9 MSI AND OTHER WAYS OF EXAMINING BIOLOGICAL TISSUES

Thanks to the wide range of ionization techniques and availability of high-resolution analyzers, the application of MS encompasses broad science areas in the fields of proteomics, metabolomics, pharmacy, toxicology, environmental applications, isotope analyses, carbon dating, or homeland security. The application of MS in industry is logical for evaluating the quality of raw materials and final products. A bright example is the petrochemical industry [61]. In biology and health sciences, it is now a standard tool for the identification, characterization, and quantification of proteins and organic compounds. MS in clinical diagnostics and toxicology progressively complements spectroscopic or immunoaffinity analytical methods [62]. MSI (also imaging mass spectrometry—IMS) is a technique which allows to investigate spatial molecular distribution in a section of the exact area of a biological sample, typically in a sliced tissue [63]. MSI can provide qualitative as well as quantitative information for a wide variety of compounds such as proteins, peptides, lipids, metabolites, drugs, and others. The final output is a two-dimensional image based on the determined molecular masses. All acquired images from consecutive tissue slices can be combined to a three-dimensional model of the examined object showing the distribution profiles of individual detected compounds [64]. MALDI MS and secondary ion mass spectrometry (SIMS) are the most typical approaches in MSI. Ambient ionization techniques such as DESI, laser ablation electrospray ionization (LAESI), or liquid extraction surface analysis (LESA) represent an alternative [65]. MALDI imaging allows a simultaneous mapping of biological molecules present in thin tissue sections placed on the MALDI probe. The principle is similar to that of scanning microscopy: the whole sample area is set for analysis, divided into closely adjacent positions for laser firing and then sequentially analyzed. Only those instruments with a gentle target displacement and fine laser beam focus can be used. The information obtained by MALDI MSI is suitably complemented by histological and immunological observations, which help to search for molecular markers [66]. The sample preparation procedure is a key step and requires a precise preparation of thin tissue slices, their washing, desalting, and the subsequent application of matrix (sublimation or spraying methods are used to evenly cover the entire surface area of the sample) [67]. Proteins can be identified directly from the sliced tissue after in situ digestions [68]. The spatial resolution ranges from 5 to 200 μm.

A higher spatial resolution of 50–100 nm can be achieved using SIMS [69]. SIMS was originally used to analyze solid surfaces with a focused primary ion beam and collecting and analyzing the released secondary ions. In biological research, SIMS is applied to study biomolecules with relatively low molecular masses. SIMS instruments work in two different modes: static and dynamic. In the dynamic SIMS, the sample is bombarded constantly by primary ions (Cs^+ or O^-), which causes an excessive analyte fragmentation. An in-depth chemical profile of the observed material can be acquired in this way. Conversely, the static SIMS uses only pulsed primary ion beams to avoid any sample surface distortion and provides the best surface-sensitive molecular resolution [69]. By combining these two SIMS approaches, the analysis by defined layers can be achieved. The first layer of a sample is explored by the soft static mode. Then the analyzed surface is eroded by more invasive atom beams which reveal a lower layer and the process is repeated.

DESI is commonly used for imaging of lipids [70]. Thin-layer chromatography probes used for the separation of organic molecules can be effectively analyzed by DESI MS [65]. By combining the benefits of ESI and MALDI, the hybrid ionization technique matrix-assisted laser desorption electrospray ionization (MALDESI) has been introduced. Here the ions are generated by a laser ablation and electrospray postionization [71]. LAESI combines a mid-infrared laser ablation with a secondary ESI process for samples containing water (liquids or tissues). The laser is tuned to the absorption line of water, but the ejected secondary material is not ionized. Therefore, an ESI source is located above the sample for a post-ablation ionization [72]. LESA represents a different attitude for sampling from tissue specimens. First, the analyte is extracted from the sample surface by a liquid microjunction between the probe and sample. Then the droplet is ionized and analyzed [73]. LESA shows a low spatial resolution, but the noninvasive sampling and easy connectivity to MS analyzers are promising for LESA-based MSI under native conditions.

In addition to MSI, several other interesting approaches for analyzing biological material have been developed on the basis of ambient ionization. Paper spray MS was invented for a fast and direct analysis of molecules such as lipid, hormones, and drugs in tissues without a complicated sample pretreatment. A sample drop (1 μL) is deposited on a triangle shaped paper directed by one of its tips to the MS inlet. For ionization, both a high voltage and solvent are applied, and the analyte is sprayed from the paper tip into the MS instrument [74]. A more convenient method for tissue analysis is the rapid evaporative ionization mass spectrometry (REIMS) invented for a medical use [75]. This electrosurgical dissection device is combined with an air pump and flexible tubing connected to the ion source. The charged molecules released during the dissection are sucked and transferred to the MS instrument and analyzed in real time. The system is marketed nowadays as "iKnife." REIMS is applicable also for the analysis of intact microorganisms [76]. Nanostructure-initiator MS (NIMS) is a desorption/ionization technique, which uses silicon fibers as laser energy absorbers. The absorption of laser energy results in a fast vaporization of the sample and its ionization. A repeated application of liquid sample on the nanostructure initiator and its drying can be used for analyte enrichment. Small organic molecules such as metabolites can be detected by this approach even in the yoctomole (10^{-24}) range [77].

2.10 MASS SPECTRUM, DATA REPRESENTATION, AND MANAGEMENT

Mass spectrometric experiments involve a sequential use of several software types. First, an acquisition software is employed for acquiring mass spectra and their saving together with metadata of the experiment. The mass spectra are then processed (e.g., with smoothing and peak picking steps) by means of a processing software. Finally, specialized software tools are applied, for example, in proteomics to perform database searches and calculating quantification results. Apart from the proprietary software provided by instrument vendors, there is also both free and open-source software available for MS. To mention a few examples, mMass has been developed for mass spectrometric data analysis and interpretation [78], MaxQuant [79], and Skyline [80] are applicable in quantitative proteomics, OpenChrom is useful for evaluating GC- or LC-MS data [81], and BIOSPEAN allows to compare intact cell MALDI-TOF mass spectra [82].

A common mass spectrum represents a plot of m/z values (on the x-axis) of all detected ions against their abundance ($\sim$ signal intensity; on the y-axis). The most abundant ion is designated a base peak. On a relative scale, abundances of the other ions are expressed as percentage abundances relative to this peak [3]. Mass spectra can be stored in two major ways: as continuous spectra ("profile-mode") or peak lists ("centroided"). Profile-mode spectra comprise spaced data points and thus each peak has a defined shape. Conversely, a peak list is represented by m/z and intensity pairs extracted from the original peaks. The latter representation saves memory space but substantial information (resolution) is lost [83].

In biological research, MS commonly produces large amounts of data. Mass-spec manufacturers as well as organized user communities have developed various formats for data storage, exchange, and processing. There are many instrument data formats (or the proprietary software generates folders with multiple files) with extensions such as .baf, .fid or .yep (Bruker), .wiff or .t2d (AB Sciex), .qgd (Shimadzu), and .raw (Thermo, Waters), which are not interchangeable and transferable. For database searches using MS/MS data for protein identification, the output files from mass spectrometers are commonly converted into simple text files. The MGF format has been launched by Matrix Science (London, United Kingdom), and it encodes multiple MS/MS spectra in a single file, which is applicable with Mascot, the most common search engine. However, many valuable metadata are lost during the respective conversion [83]. To cope with these disadvantages, attempts appeared in 2003 to introduce a standardized MS data format. The Human Proteome organization (HUPO) has made a big effort via its Proteomics Standards Initiative (PSI) to manage the development of open data formats [83]. All it has started with a requirement to keep most of the information from each experimental run that would easily be accessible by any tool. First, mzData was developed by the PSI itself. Independently, the Institute for Systems Biology (Seattle, The United States) produced mzXML [84]. Because of an inconvenience of having two open formats for preserving the same information, the two institutions have created mzML as a compromise and new format with best features from both mzXML and mzData [85]. For quantitative proteomics analyses, mzQuantML and mzTab formats are currently being developed by the HUPO PSI (http://www.psidev.info).

ACKNOWLEDGMENTS

This work was supported by grant No. LO1204 (National Program of Sustainability I) from the Ministry of Education, Youth and Sports, Czech Republic.

REFERENCES

[1] Price P. Standard definitions of terms relating to mass spectrometry. A report from the Committee on Measurements and Standards of the American Society for Mass Spectrometry. J Am Soc Mass Spectrom 1991;2:336—48.

[2] Cooks RG, Rockwood AL. The "Thomson": suggested unit for mass spectroscopists,. Rapid Commun Mass Spectrom 1991;5:93.

[3] Dass C. Basics of mass spectrometry. In: Fundamentals of contemporary mass spectrometry. Hoboken, NJ: John Wiley & Sons, Inc; 2007. p. 3—14.

[4] Lössl P, M. van de Waterbeemd, Heck AJ. The diverse and expanding role of mass spectrometry in structural and molecular biology. EMBO J 2016;35:2634—57.

[5] Moore CE, Jaselskis B, von Smolinski AJ. The proton. J Chem Educ 1985;62:859—60.

[6] Hedenus M. Eugen Goldstein and his laboratory at Berlin Observatory. Astron Nachr 2002;6:567—9.

[7] Rüchardt E. Zur Entdeckung der Kanalstrahlen vor fünfzig Jahren. Naturwissenschaften 1936;24:57—62.

[8] Thomson JJ. Rays of positive electricity and their application to chemical analyses. In: Thomson JJ, Horton F, editors. Monographs on physics. London: Longmans, Green & Co.; 1913. p. 1—132. Reprinted by the American Society for Mass Spectrometry in 2013.

[9] Thomson JJ. Rays of positive electricity. Proc Roy Soc London 1913;A89:1—20.

[10] Squires G. Francis Aston and the mass spectrograph. Dalton Trans 1998;23:3893—900.

[11] Dempster AJ. A new method of positive ray analysis. Phys Rev 1918;11:316—25.

[12] Allison SK. Arthur Jeffrey Dempster 1886-1950. Biogr Mems Nat Acad Sci 1952;319—33.

[13] Wood HG, Werkman CH, Hemingway A, Nier AO. Heavy carbon as a tracer in bacterial fixation of carbon dioxide. J Biol Chem 1940;135:789—90.

[14] Meyerson S. Reminiscence of the early days of mass spectrometry in the petroleum industry. Org Mass Spectrom 1986;21:197—208.

[15] Paul W, Steinwedel H. Ein neues Massenspektrometer ohne Magnetfeld. Zeitschrift für Naturforschung A 1953;8:448—50.

[16] Jennings KR. Collision-induced decompositions of aromatic molecular ions. Int J Mass Spectrom Ion Phys 1968;1:227—35.

[17] Barber M, Bordoli RS, Sedgwick RD, Tyler AN. Fast atom bombardment of solids as an ion source in mass spectrometry. Nature 1981;293:270—5.

[18] Fenn JB, Mann M, Meng CK, Wong SF, Whitehouse CM. Electrospray ionization for mass spectrometry of large biomolecules. Science 1989;246:64—71.

[19] Karas M, Bachmann D, Bahr U, Hillenkamp F. Matrix-assisted ultraviolet laser desorption of non-volatile compounds. Int J Mass Spectrom Ion Processes 1987;78:53—68.

[20] Tanaka K, Waki H, Ido Y, Akita S, Yoshida Y, Yoshida T. Protein and polymer analyses up to m/z 100 000 by laser ionization time-of-flight mass spectrometry. Rapid Commun Mass Spectrom 1988;2:151—3.

[21] Takats Z, Wiseman JM, Gologan B, Cooks RG. Mass spectrometry sampling under ambient conditions with desorption electrospray ionization. Science 2004;306:471—3.

[22] Makarov A. Electrostatic axially harmonic orbital trapping: high-performance technique of mass analysis. Anal Chem 2000;72:1156—62.

[23] Vékey K. Internal energy effects in mass spectrometry. J Mass Spectrom 1996;31:445—63.

[24] El-Aneed A, Cohen A, Banoub J. Mass spectrometry, review of the basics: electrospray, MALDI, and commonly used mass analyzers. Appl Spectrosc Rev 2009;44:210−30.

[25] Dass C. Mass analysis and ion detection. In: Fundamentals of contemporary mass spectrometry. Hoboken, NJ: John Wiley & Sons, Inc; 2007. p. 67−117.

[26] Kopenaal DW, Barinaga CJ, Denton MB, Sperline RP, Hieftje GM, Schilling GD, et al. MS detectors. Anal Chem 2005;77:418A−27A.

[27] Gross JH. Instrumentation. In: Mass spectrometry: a textbook. 3rd Edition Cham, Switzerland: Springer International Publishing AG; 2017. p. 151−292.

[28] Monge ME, Harris GA, Dwivedi P, Fernández FM. Mass spectrometry: recent advances in direct open air surface sampling/ionization. Chem Rev 2013;113:2269−308.

[29] Aebersold R, Mann M. Mass spectrometry-based proteomics. Nature 2003;422:198−207.

[30] Bruins AP. Mechanistic aspects of electrospray ionization. J Chromatogr A 1998;794:345−57.

[31] Bruins AP. Atmospheric-pressure-ionization mass spectrometry: I. Instrumentation and ionization techniques. Trac-Trends Anal Chem 1994;13:37−43.

[32] Raffaelli A, Saba A. Atmospheric pressure photoionization mass spectrometry. Mass Spectrom Rev 2003;22:318−31.

[33] Zenobi R, Knochenmuss R. Ion formation in MALDI mass spectrometry. Mass Spectrom Rev 1998;17:337−66.

[34] Dreisewerd K. Recent methodological advances in MALDI mass spectrometry. Anal Bioanal Chem 2014;406:2261−78.

[35] Cooks RG, Ouyang Z, Takats Z, Wiseman JM. Detection technologies. Ambient mass spectrometry. Science 2006;311:1566−70.

[36] Wiseman JM, Puolitaival SM, Takáts Z, Cooks RG, Caprioli RM. Mass spectrometric profiling of intact biological tissue by using desorption electrospray ionization. Angew Chem Int Ed 2005;44:7094−7.

[37] Cody RB, Laramée JA, Durst HD. Versatile new ion source for the analysis of materials in open air under ambient conditions. Anal Chem 2005;77:2297−302.

[38] Nier AO. The development of a high resolution mass spectrometer: a reminiscence. J Am Soc Mass Spectrom 1991;2:447−52.

[39] Chernushevich IV, Loboda AV, Thomson B. An introduction to quadrupole−time-of-flight mass spectrometry. J Mass Spectrom 2001;36:849−65.

[40] Wollnik H. History of mass measurements in time-of-flight mass analyzers. Int J Mass Spectrom 2013;349-350:38−46.

[41] March RE. An introduction to quadrupole ion trap mass spectrometry. J Mass Spectrom 1997;32:351−69.

[42] Schwarz JC, Senko MW, Syka JEP. A two-dimensional quadrupole ion trap mass spectrometer. J Am Soc Mass Spectrom 2002;13:659−69.

[43] Marshall AG, Hendrickson CL, Jackson GS. Fourier transform ion cyclotron resonance mass spectrometry: a primer. Mass Spectrom Rev 1998;17:1−35.

[44] Scigelova M, Makarov A. Orbitrap mass analyzer—overview and applications in proteomics. Proteomics 2006;6(Suppl. 2):16−21.

[45] de Hoffmann E. Tandem mass spectrometry: a primer. J Mass Spectrom 1996;31:129−37.

[46] Yost RA, Enke CG. Selected ion fragmentation with a tandem quadrupole mass spectrometer. J Am Chem Soc 1978;100:2274−5.

[47] Cottrell JS. Protein identification using MS/MS data. J Proteomics 2011;74 1842−1851.

[48] Dass C. Tandem mass spectrometry. In: Fundamentals of contemporary mass spectrometry. Hoboken, NJ: John Wiley & Sons, Inc; 2007. p. 119−50.

[49] Domon B, Aebersold R. Mass spectrometry and protein analysis. Science 2006;312:212−17.

[50] Picotti P, Aebersold R. Selected reaction monitoring−based proteomics: workflows, potential, pitfalls and future directions. Nat Methods 2012;9:555−66.

[51] Bourmaud A, Gallien S, Domon B. Parallel reaction monitoring using quadrupole-Orbitrap mass spectrometer: principle and applications. Proteomics 2016;16:2146−59.

[52] Sleno L, Volmer DA. Ion activation methods for tandem mass spectrometry. J Mass Spectrom 2004;39:1091−112.

[53] Tomer KB. Separations combined with mass spectrometry. Chem Rev 2001;101:297−328.

[54] Dass C. Hyphenated separation techniques. In: Fundamentals of contemporary mass spectrometry. Hoboken, NJ: John Wiley & Sons, Inc; 2007. p. 151−94.

[55] Abian J. The coupling of gas and liquid chromatography with mass spectrometry. J Mass Spectrom 1999;34:157−68.

[56] Holčapek M, Jirásko R, Lísa M. Recent developments in liquid chromatography−mass spectrometry and related techniques. J Chromatogr 2012;A 1259:3−15.

[57] Mukhopadhyay R. The automated union of LC and MALDI MS. Anal Chem 2005;77:150A−2A.

[58] Issaq HJ. The role of separation science in proteomics research. Electrophoresis 2001;22:3629−38.

[59] Schmitt-Kopplin F, Frommberger M. Capillary electrophoresis−mass spectrometry: 15 years of developments and applications. Electrophoresis 2003;24:3837−67.

[60] Kostal V, Katzenmeyer J, Arriaga EA. Capillary electrophoresis in bioanalysis. Anal Chem 2008;80.4533 50.

[61] Marshall AG, Rodgers RP. Petroleomics: chemistry of the underworld. Proc Natl Acad Sci USA 2008;105:18090−5.

[62] Vogeser M. Mass spectrometry in the clinical laboratory—challenges for quality assurance. Spectroscopy 2015;13:14−19.

[63] McDonnell LA, Heeren RMA. Imaging mass spectrometry. Mass Spectrom Rev 2007;26:606−43.

[64] Schwamborn K, Caprioli RM. MALDI imaging mass spectrometry—painting molecular pictures. Mol Oncol 2010;4:529−38.

[65] Wu C, Dill AL, Eberlin LS, Cooks RG, Ifa DR. Mass spectrometry imaging under ambient conditions. Mass Spectrom Rev 2012;32:218−43.

[66] Chughtai K, Heeren RMA. Mass spectrometric imaging for biomedical tissue analysis. Chem Rev 2010;110:3237−77.

[67] Chaurand P, Schwartz SA, Reyzer ML, Caprioli RM. Imaging mass spectrometry: principles and potentials. Toxicol Pathol 2005;33:92−101.

[68] Balluff B, Rauser S, Ebert MP, Siveke JT, Höfler H, Walch A. Direct molecular tissue analysis by MALDI imaging mass spectrometry in the field of gastrointestinal disease. Gastroenterology 2012;143:544−9.

[69] Gamble LJ, Anderton CR. Secondary ion mass spectrometry imaging of tissues, cells, and microbial systems. Micros Today 2016;24:24−31.

[70] Ferguson CN, Fowler JWM, Waxer JF, Gatti RA, Loo JA. Mass spectrometry-based tissue imaging of small molecules. advances in experimental medicine and biology. In: Woods A, Darie C, editors. Advancements of Mass Spectrometry in Biomedical Research Advances in Experimental Medicine and Biology, vol 806. Cham, Switzerland: Springer; 2014. p. 283−99.

[71] Robichaud G, Barry JA, Muddiman DC. IR-MALDESI mass spectrometry imaging of biological tissue sections using ice as a matrix. J Am Soc Mass Spectrom 2014;25:319−28.

[72] Nemes P, Vertes A. Laser ablation electrospray ionization for atmospheric pressure, in vivo, and imaging mass spectrometry. Anal Chem 2007;79:8098−106.

[73] Griffiths RL, Creese A, Race AM, Bunch J, Cooper HJ. LESA FAIMS mass spectrometry for the spatial profiling of proteins from tissue. Anal Chem 2016;88:6758−66.

[74] Wang H, Manicke NE, Yang Q, Zheng L, Shi R, Cooks RG, et al. Direct analysis of biological tissue by paper spray mass spectrometry. Anal Chem 2011;83:1197−201.

[75] Balog J, Szaniszlo T, Schaefer KC, Denes J, Lopata A, Godorhazy L, et al. Identification of biological tissues by rapid evaporative ionization mass spectrometry. Anal Chem 2010;82:7343−50.

[76] Strittmatter N, Jones EA, Veselkov KA, Rebec M, Bundy JG, Takats Z. Analysis of intact bacteria using rapid evaporative ionisation mass spectrometry. Chem Commun 2013;49:6188−90.

[77] Greving MP, Patti GJ, Siuzdak G. Nanostructure-initiator mass spectrometry metabolite analysis and imaging. Anal Chem 2011;83:2−7.

[78] Strohalm M, Hassman M, Košata B, Kodíček M. mMass data miner: an open source alternative for mass spectrometric data analysis. Rapid Commun Mass Spectrom 2008;22:905−8.

[79] Cox J, Mann M. MaxQuant enables high peptide identification rates, individualized p.p.b.-range mass accuracies and proteome-wide protein quantification. Nat Biotechnol 2008;26:1367−72.

[80] MacLean B, Tomazela DM, Shulman N, Chambers M, Finney GL, Frewen B, et al. Skyline: an open source document editor for creating and analyzing targeted proteomics experiments. Bioinformatics 2010;26:966−8.

[81] Wenig P, Odermatt J. OpenChrom: a cross-platform open source software for the mass spectrometric analysis of chromatographic data. BMC Bioinform 2010;11:405.

[82] Raus M, Šebela M. BIOSPEAN: a freeware tool for processing spectra from MALDI intact cell/spore mass spectrometry. J Proteomics Bioinform 2013;6:283−7.

[83] Deutsch EW. File formats commonly used in mass spectrometry proteomics. Mol Cell Proteomics 2012;11:1612−21.

[84] Pedrioli PG, Eng JK, Hubley R, Vogelzang M, Deutsch EW, Raught B, et al. A common open representation of mass spectrometry data and its application to proteomics research. Nat Biotechnol 2004;22:1459−66.

[85] Martens L, Chambers M, Sturm M, Kessner D, Levander F, Shofstahl J, et al. mzML—a community standard for mass spectrometry data. Mol Cell Proteomics 2011;10;R110.000133.

CHAPTER THREE

MALDI-TOF Commercial Platforms for Bacterial Identification

Antonio Sampedro[1], Julián Ceballos Mendiola[1] and Luís Aliaga Martínez[2]

[1]Department of Microbiology, Hospital Virgen de las Nieves, Granada, Spain
[2]Internal Medicine Department, University Hospital Health Campus, Granada, Spain

3.1 INTRODUCTION

Mass spectrometry (MS) technology has been used for several decades in chemistry, but the concept of using MS to identify bacteria was proposed in the 1970s [1]. In the 1980s, desorption/ionization techniques that allow the generation of molecular biomarker ions from microorganisms were developed and whole proteomics became possible [2,3].

In the late 1980s, Tanaka and Fenn, thanks to the development of soft ionization techniques (matrix-assisted laser desorption/ionization (MALDI) and electrospray ionization), made possible the analysis of large biomolecules such as intact proteins [3,4].

In 1996, MALDI-time of flight (TOF) spectral fingerprints could be obtained from whole bacterial cells [5]. In the 1990s libraries of reference spectra and software for bacterial identification were developed, and this allowed its commercialization. Now, MALDI-TOF MS has emerged as a rapid, accurate, and sensitive tool for microbial characterization and identification of bacteria, fungi, and viruses and being rapidly adopted worldwide.

3.2 COMMERCIAL MALDI-TOF MS SYSTEMS

MALDI-TOF system platforms are composed of MALDI-TOF mass spectrometer, a licensed software, which is utilized by users for the

The Use of Mass Spectrometry Technology (MALDI-TOF) in Clinical Microbiology.
DOI: https://doi.org/10.1016/B978-0-12-814451-0.00003-4

acquisition and saving of mass spectra, processing of the acquired spectra, spectra comparison, and searches against spectral or sequence databases. Each MALDI-TOF mass spectrometer is composed of three principal units. The first is the ion source that makes ionization possible and transfers sample molecule ions into a gas phase. The second unit is the mass analyzer that allows ion separation according to mass to charge ratio (m/z). The last unit is a detection device for monitoring separated ions.

Currently two MALDI-TOF MS systems are available: MALDI Biotyper System (Bruker Daltonics, Heidelberg, Germany) and VITEK MS (bioMérieux, Marcy l'Etoile, France). In addition to Bruker and bioMérieux, the Andromas software and database for microbial identification using MALDI-TOF MS [6] has been offered since 2012 in Europe.

Andromas, VITEK MS, and MALDI Biotyper have been accredited for identification purposes in clinical microbiology laboratories under EU directive EC/98/79 in several European countries. The VITEK MS and the MALDI Biotyper System were cleared by the US Food and Drug Administration (FDA) for the identification of cultured bacteria and yeast.

Spectral databases are often marketed as part of a proprietary system and are constructed and maintained by their manufacturers. As each proprietary system uses its own algorithms, databases, software, and interpretive criteria for microbial identification, numerical data (i.e., spectral scores) from different commercial systems are not directly comparable [7]. Bruker's Microflex LT mass spectrometer is a desktop instrument, whereas BioMérieux is a larger, floor model.

In this chapter the three systems will be commented, although the performance of the Biotyper and VITEK MS will be compared.

3.3 MALDI BIOTYPER SYSTEM

The MALDI Biotyper was the first MALDI-TOF MS system capable of microbial identification. The system consisted of a basic MALDI-TOF platform (benchtop Microflex LT/SH mass spectrometer or the high-end Microflex LT/SH smart), a disposable MBT Biotarget 96 or a reusable 48-spot MALDI target plate, operating and analysis software (MALDI Biotyper software), an onsite database and a simple method for extraction/preparation of sample [8].

Bruker's system is the most studied in terms of publications and evaluations in multiple laboratories. Reference mass spectra are generated from a set of about 20 raw spectra derived from a single sample of a unique strain of a microorganism. The reference mass spectra consist of the most prevalent peaks in the set of the 20 raw spectra characterized by the average mass and the average intensity of peaks. About 70 peaks with a minimum frequency of 25% of a peak within the about 20 spectrum set compose the reference mass spectra. [9]. The Biotyper software allows the user to create main spectra with microorganisms isolated in the laboratory.

For microorganisms' identification, spectra are generated and are analyzed with respect to the frequency, position, and intensity of the peaks in the spectrum. The unknown spectrum is converted into a mass peak list and their corresponding intensities. Then, these spectra are compared with the main spectra in the Biotyper database. The Biotyper system assigns a score between zero and three to the unknown spectrum according to the comparison of unknown peak list with the reference mass spectra peak list of the database. The results of MALDI Biotyper identification are based on the interpretation of the log(score) of the best match. According to the manufacturer, a log(score) > 2.3 indicates "highly probable species identification," a log(score) > 2 and < 2.299 indicates "secure genus identification, probable species identification," log (score) > 1.7 and < 1.999 indicates "probable genus identification," and log(score) < 1.7 indicates "nonreliable identification."

The library IVD MALDI Biotyper (version CE-marked according to the European Union for In Vitro Diagnostic) covers a broad range of microorganisms including nonfermenting Gram-negative bacteria, *Enterobacteriaceae*, other Gram-negative bacteria, Gram-positive bacteria, and yeasts. In 2015 MALDI Biotyper got IVD certification for MALDI Sepsityper Kit for identification of microorganisms directly from positive blood culture and mycobacteria identification. The MBT Mycobacteria Library covers 164 of the currently known 180 species, with 780 strains. Currently MALDI Biotyper library consists of >6900 strains.

In 2013 the MALDI Biotyper CA System has been granted the US FDA clearance for the identification of Gram-negative bacterial colonies cultured from human specimens with a subsequent extension to libraries containing aerobic Gram-positive, fastidious Gram-negatives, *Enterobacteriaceae*, anaerobic bacteria and yeasts. In 2016 new microorganism (molds, mycobacteria, and Nocardia) are added to the already cleared device (https://www.accessdata.fda.gov/cdrh_docs/pdf16/ K163536.pdf).

3.4 VITEK MS SYSTEM

The SARAMIS (Spectral Archiving And Microbial Identification System) software with database was created by AnagnosTec GmbH (Zossen, Germany) and marketed by Shimadzu together with Axima mass spectrometers (Shimadzu, Columbia, MD), before being acquired by bioMérieux in 2010 for its incorporation to the VITEK MS platform [7].

The VITEK MS v3.0 system (bioMérieux, Marcy l'Etoile, France) is a system consisting of VITEK MS-DS target slides, VITEK MS Prep Station, Knowledge Base v3.0, software, and the VITEK MS (original equipment manufacturer Shimadzu AXIMA mass assurance spectrometer).

VITEK MS V3 CE marked and FDA cleared database for bacteria and fungi and contains mycobacteria, Nocardia, and molds. This database includes mycobacteria tuberculosis, non-tuberculous mycobacteria, dermatophytes and dimorphic fungi, including the most prevalent species relevant to patient infections (https://www.accessdata.fda.gov/cdrh_docs/ reviews/K162950.pdf).

The reference database for the VITEK MS system includes data representing 1046 taxa, including 882 bacteria and 164 fungi.

The creation of a reference mass spectrum is based on a "mass binning" algorithm. The preprocessed spectrum is divided into 1300 predefined intervals on a mass range from 3000 to 17,000 Da. These intervals are called "bins." Only the peak with the highest intensity is retained in each bin, and the other peaks are discarded. This algorithm transforms a long list of peaks of variable intensity into an immovable list of bins with a corresponding intensity. Next, each peak is weighted based on its specificity at the species level, the genus level, or other taxonomic level, as performed for the SARAMIS database. This algorithm is called "Advanced Spectra classifier." Each SuperSpectra of the database is constructed with the spectral information of numerous strains of the same species. Thus, the reference database contains only a matrix of 1300 bins with species-specific weights. Each species or species group is represented by an average of 17 isolates. In order to capture the degree of acceptable variation within spectra from the same species, each reference isolate was grown on multiple media types under several growth conditions. The raw spectra were then acquired by more than one technician using multiple instruments. This process resulted in an average of 36 reference spectra per species.

An unknown spectrum acquired by the mass spectrometer undergoes the same pretreatment procedure as the reference spectrum. The bin scores of the unknown spectrum are multiplied with the weight of each species-specific weighted-bin included in the reference matrix. The sum of the weighted bins is calculated for each species contained in the reference matrix, and a confidence value of the unknown spectrum is calculated for each species included in the matrix [9]. The output is a confidence value, ranging from 0% to 100%. Values between 60% and 99% give a correct identification and values close to 99.9% indicate a closer approximation exact; when the probability percentage is less than 60% the microorganism is considered as not identified [10].

3.5 MALDI BIOTYPER VS VITEK MS SYSTEMS' COMPARISON

Comparison of the Bruker Biotyper and VITEK MS has demonstrated similar performance for the identification of bacterial isolates. While some studies have found the Bruker Biotyper to provide a higher proportion of good identifications (94.4% vs 88.8%) for all isolates tested and among nonfermenters (97.0% vs 89.5%), other studies have demonstrated better performance of the VITEK MS for identification of anaerobes and viridans group streptococci [11–14].

Jamal et al. compared MALDI Biotyper vs VITEK MS using 806 pathogens comprising 507 Gram-negative bacilli, 16 Gram-negative cocci, 267 Gram-positive cocci, and 16 Gram-positive bacilli; Bruker System and VITEK MS correctly identified isolates to genus and species levels 97.3% and 93.2%, and 99.8% and 99.0%, respectively. They report that both systems correctly identified the majority of the species in the family *Enterobacteriaceae, Pseudomonas* spp., and *Acinetobacter baumannii* [15].

Bruker System performs better than automated biochemical identification for commonly yeast and bacteria and better for unusual organisms. Saffert et al. compared the BD Phoenix System and Bruker Biotyper using 440 Gram-negative bacilli (common and infrequent) isolates in the clinical laboratory, using biochemical testing and/or genetic sequencing to resolve discordant results; Bruker Biotyper correctly identified 93% and 82% of isolates to the genus and species levels, respectively, while BD Phoenix correctly identified 83% and 75%, respectively. These differences were greater for the infrequent isolates [16].

VITEK MS performs as well as MALDI Biotyper. In a study related by Martiny et al. of 1129 routine bacterial isolates (including anaerobes and enteropathogens), they report 93% species-level correctly identified with VITES MS v 1.1 database [14]. Richter et al. showed that VITEK MS correctly identified 97% of the *Enterobacteriaceae* to the genus level and 84% to the species level [17].

With regard to fastidious Gram-negative bacteria, Branda et al. found that VITEK MS identified 97% and 96% of 226 isolates of fastidious Gram-negative bacteria to the genus and species levels, respectively [18].

In another study to evaluate Bruker Biotyper for Gram-positive isolates such as *Staphylococci*, *Streptococci*, and *Enterococci*, the correct identification at the species and genus level was 98% and 79%, respectively [19]. Karpanoja et al. compared the MALDI Biotyper and VITEK MS IVD systems for the identification of viridans group *Streptococci* using 54 type strains and 97 blood culture isolates [20]. The Biotyper and VITEK MS systems yielded correct species-level identification for 94% and 69% of the type strains and correctly classified 89% and 93% of the blood culture isolates to the group level, respectively. Both systems have problems for viridans group *Streptococci* and *Streptococcus pneumoniae* identification due to the fact that *S. pneumoniae* is strikingly similar to viridans *Streptococci* genetically. In contrast, another study did not misidentify any of the 369 non-pneumococcal *Streptococci* and related genera as *S. pneumoniae* [21]. Then MALDI-TOF MS analysis cannot be used alone for the identification of these organisms.

There are several studies to evaluate VITEK MS and Bruker System for the identification of anaerobic bacteria. Barreau et al. analyzed 1325 anaerobes using MALDI Biotyper (v3.0 software) and showed that 100% and 93% of the isolates were correctly identified at the genus and species levels, respectively [22]. On the other hand, Garner et al. evaluated the VITEK MS system (database v2.0) using 651 anaerobic isolates and reported that 93% and 91% of anaerobic bacteria were correctly identified to the genus and species levels, respectively [23].

In addition to bacteria, MALDI Biotyper and VITEK MS can identify yeasts. The performance of MALDI-TOF MS in yeast identification is good with identification rates above 90% in most studies.

The different studies that evaluate Bruker Biotyper for yeast identification used a complete extraction using formic acid and acetonitrile. Dhiman et al. evaluated the Bruker Biotyper system for identification of 241 common and uncommon yeasts finding a 96% of correct identification at species level for common and 85% for infrequent yeast [24]. These results are comparable with those obtained by Marklein et al. who reported 92.5% and Stevenson et al. reported 87.1% species-level identification using a Bruker System [25,26]. More recently, Lacroix et al. analyzed 1383 *Candida* isolates using Bruker System and showed that 98.3% of the isolates were correctly identified at the species level vs 96.5% by conventional techniques; the major drawback of the Bruker Biotyper is found in the identification of rare yeast species such as *Candida fabiani*, *Candida fermentati*, and *Candida ethanolica* due to a lack of reference spectra in database [27].

Several studies investigated VITEK MS for yeast identification performance, and most of them employ formic acid extraction. The correct identification rates ranged from 84.3% [28] to 97.5% [29]; the major drawback of VITEK MS is in the identification of *Candida rugosa*, *Candida dubliniensis* and *Candida parapsilosis*, and *Cryptococcus* species [28,30,31].

Molds identification using MALDI-TOF MS has presented more difficulties than that of bacteria and yeasts analysis, due to the requirement for sample preparation (VITEK MS and MALDI Biotyper includes an extraction step with acetonitrile) and the variability of spectra according to the growing conditions. The species coverage of MALDI reference databases constitutes the weakness of MALDI-TOF systems and may not be sufficient for routine identification. The employment of in-house databases for molds (in combination with MALDI-TOF library) can improve for the identification of any species of clinically relevant mold [32].

In filamentous fungi, the MALDI Biotyper using an in-house database correctly identifies between 72.0% and 88.9% of the isolates [9]. More recently, Riat et al., by using Bruker Biotyper software and MALDI Biotyper database (Filamentous Fungi Library 1.0), found that 91.6% of mold isolates had a cutoff value >1.7 [33].

In a large evaluation study of VITEK MS version 3.0, carried out with 319 mold isolates from 43 different genera, recovered from different clinical specimens, the MALDI-TOF MS system was able to correctly identify 66.8% (213/319) of isolates. This method was able to accurately identify 133/144 (93.6%) *Aspergillus* sp. isolates to the species level; in contrast, only 65% (17/26) of isolates of *Fusarium* species were identified. Of the unidentified isolates, 69 were not in the database [34].

3.6 ANDROMAS

The Andromas system is a database manufactured and maintained by Andromas SA (Paris, France). The software separates entries into three separate databases dedicated to the identification of bacteria, mycobacteria, yeasts, and *Aspergillus* spp.

Each entry has been chosen for its medical interest and, in most cases, was obtained from a clinical strain identified by molecular biology.

Reference spectra are created from a set of 10 raw spectra acquired from 10 subcultures of a strain. Each raw spectrum is first preprocessed. The intensity of each peak is then transformed in a ratio (peak intensity/intensity of the most intense peak). Only peaks with a relative intensity superior to 0.1 and present in all 10 spectra are conserved and constitute the reference spectra. To increase the robustness of this method, the system uses multiple species-specific spectral profiles for each organism in the database. These spectra could correspond either to profiles of the same strain of bacterium grown on different types of culture media [9].

The unknown raw spectrum is preprocessed and converted into a mass peak list, in which only peaks with relative intensity above 0.1 are conserved. The unknown spectrum is then compared with each reference spectra included in the reference database.

The results of the Andromas software are given as a percentage of similarity between the spectrum generated by the microorganism and a database reference spectrum. Identification are grouped in three categories: "good identification" (the similarity percent is superior or equal to 65% with a difference between the first two hits between different species of at least 10%), "identification to be confirmed" (the similarity percent ranges from 60% to 64%), and "no identification" (the similarity percent is under 60% or the difference between the first two hits between different species is above 10%) [6].

REFERENCES

[1] Anhalt JP, Fenselau C. Identification of bacteria using mass spectrometry. Anal Chem 1975;47:219–25.

[2] Karas M, Hillenkamp F. Laser desorption ionization of proteins with molecular masses exceeding 10,000 daltons. Anal Chem 1988;60:2299–301.

[3] Tanaka K, Hiroaki W, Ido Y, Akita S, Yoshida Y, Yoshida T. Protein and polymer analyses up to m/z 100 000 by laser ionization time-of-flight mass spectrometry. Rapid Commun Mass Spectrom 1988;2:151–253.

[4] Fenn JB, Mann M, Meng CK, Wong SF, Whitehouse CM. Electrospray ionization–principles and practice. Mass Spectrom Rev 1990;9:37–70.

[5] Holland RD, Wilkes JG, Rafii F, Sutherland JB, Persons CC, Voorhees KJ, et al. Rapid identification of intact whole bacteria based on spectral patterns using matrix-assisted laser desorption/ionization with time-of-flight mass spectrometry. Rapid Commun Mass Spectrom 1996;10:1227–32.

[6] Bille E, Dauphin B, Leto J, Bougnoux ME, Beretti JL, Lotz A, et al. MALDI-TOF MS Andromas strategy for the routine identification of bacteria, mycobacteria, yeasts, Aspergillus spp. and positive blood cultures. Clin Microbiol Infect 2012;18:17–25.

[7] Patel R. Matrix-assisted laser desorption ionization–time of flight mass spectrometry in clinical microbiology. Clin Infect Dis 2013;57:564–72.

[8] Seng P, Drancourt M, Gouriet F, La Scola B, Fournier PE, Rolain JM, et al. Ongoing revolution in bacteriology: routine identification of bacteria by matrix-assisted laser desorption ionization time-of-flight mass spectrometry. Clin Infect Dis 2009;49:543–51.

[9] Cassagne C, Normand AC, L'Ollivier C, Ranque S, Piarroux R. Performance of MALDI-TOF MS platforms for fungal identification. Mycoses 2016;59:678–90.

[10] Bilecen K, Yaman G, Ciftci U, Laleli YR. Performances and reliability of Bruker Microflex LT and VITEK MS MALDI-TOF mass spectrometry systems for the identification of clinical microorganisms. Biomed Res Int 2015;2015:516410.

[11] Cherkaoui A, Hibbs J, Emonet S, Tangomo M, Girard M, Francois P, et al. Comparison of two matrix-assisted laser desorption ionization-time of flight mass spectrometry methods with conventional phenotypic identification for routine Identification of bacteria to the species level. J Clin Microbiol 2010;48:1169–75.

[12] Jamal WY, Shahin M, Rotimi VO. Comparison of two matrix-assisted laser desorption ionization-time of flight (MALDI-TOF) mass spectrometry methods and API 20AN for identification of clinically relevant anaerobic bacteria. J Med Microbiol 2013;62:540–4.

[13] Marko DC, Saffert RT, Cunningham SA, Hyman J, Walsh J, Arbefeville S, et al. Evaluation of the Bruker Biotyper and VITEK MS matrix-assisted laser desorption ionization-time of flight mass spectrometry systems for identification of nonfermenting gram-negative bacilli isolated from cultures from cystic fibrosis patients. J Clin Microbiol 2012;50:2034–9.

[14] Martiny D, Busson L, Wybo I, El Haj RA, Dediste A, Vandenberg O. Comparison of the Microflex LT and VITEK MS systems for routine identification of bacteria by matrix-assisted laser desorption ionization-time of flight mass spectrometry. J Clin Microbiol 2012;50:1313–25.

[15] Jamal W, Albert MJ, Rotimi VO. Real-time comparative evaluation of bioMerieux VITEK MS versus Bruker Microflex MS, two matrix-assisted laser desorption-ionization time-of-flight mass spectrometry systems, for identification of clinically significant bacteria. BMC Microbiol 2014;3014:289.

[16] Saffert RT, Cunningham SA, Ihde SM, Jobe KE, Mandrekar J, Patel R. Comparison of Bruker Biotyper matrix-assisted laser desorption ionization-time of flight mass

spectrometer to BD Phoenix automated micro-biology system for identification of gram-negative bacilli. J Clin Microbiol 2011;49:887−92.

[17] Richter SS, Sercia L, Branda JA, Burnham CA, Bythrow M, Ferraro MJ, et al. Identification of *Enterobacteriaceae* by matrix-assisted laser desorption/ionization time-of-flight mass spectrometry using the VITEK MS system. Eur J Clin Microbiol Infect Dis 2013;32:1571−8.

[18] Branda JA, Rychert J, Burnham CA, Bythrow M, Garner OB, Ginocchio CC, et al. Multicenter validation of the VITEK MSv2.0 MALDI-TOF mass spectrometry system for the identification of fastidious gram negative bacteria. Diagn Microbiol Infect Dis 2014;78:129−31.

[19] Alatoom AA, Cunningham SA, Ihde SM, Mandrekar J, Patel R. Comparison of direct colony method versus extraction method for identification of gram-positive cocci by use of Bruker Biotyper matrix-assisted laser desorption ionization-time of flight mass spectrometry. J Clin Microbiol 2011;49:2868−73.

[20] Karpanoja P, Harju I, Rantakokko-Jalava K, Haanpera M, Sarkkinen H. Evaluation of two matrix-assisted laser desorption ionization-time of flight mass spectrometry systems for identification of viridans group streptococci. Eur J Clin Microbiol Infectious Dis 2014;33:779−88.

[21] Dubois D, Segonds C, Prere M-F, Marty N, Oswald E. Identification of clinical Streptococcus pneumoniae isolates among other alpha and nonhemolytic streptococci by use of the VITEK MS matrix-assisted laser desorption ionization-time of flight mass spectrometry system. J Clin Microbiol 2013;51:1861−7. Available from: https://doi.org/10.1128/JCM.03069-12.

[22] Barreau M, Pagnier I, La Scola B. Improving the identification of anaerobes in the clinical microbiology laboratory through MALDI-TOF mass spectrometry. Anaerobe 2013;2013(22):123−8.

[23] Garner O, Moción A, Branda J, Burnham CA, Bythrow M, Ferraro M, et al. Multi-centre evaluation of mass spectrometric identification of anaerobic bacteria using the VITEK® MS system. Clin Microbiol Infect 2014;20:335−9.

[24] Dhiman N, Hall L, Wohlfiel SL, Buckwalter SP, Wengenack NL. Performance and cost analysis of matrix-assisted laser desorption ionization-time of flight mass spectrometry for routine identification of yeast. J Clin Microbiol 2011;49:1614−16.

[25] Marklein G, Josten M, Klanke U, Müller E, Horré R, Maier T, et al. Matrix-assisted laser desorption ionization−time of flight mass spectrometry for fast and reliable identification of clinical yeast isolates. J Clin Microbiol 2009;47:2912−17.

[26] Stevenson LG, Drake SK, Shea YR, Zelazny AM, Murray PR. Evaluation of matrix-assisted laser desorption/ionization time-of-flight mass spectrometry (MALDI-TOF) for the identification of clinically important yeast species. J Clin Microbiol 2010;48:3482−6.

[27] Lacroix C, Gicquel A, Sendid B, Meyer J, Accoceberry I, François N, et al. Evaluation of two matrix-assisted laser desorption ionization-time of flight mass spectrometry (MALDI-TOF MS) systems for the identification of Candida species. Clin Microbiol Infect 2014;20:153−8.

[28] Mancini N, De Carolis E, Infurnari L, Vella A, Clementi N, Vaccaro L, et al. Comparative evaluation of the Bruker Biotyper and VITEK MS matrix-assisted laser desorption ionization-time of flight (MALDI-TOF) mass spectrometry systems for identification of yeasts of medical importance. J Clin Microbiol 2013;51:2453−7.

[29] Wang W, Xi H, Huang M, Wang J, Fan M, Chen Y, et al. Performance of mass spectrometric identification of bacteria and yeasts routinely isolated in a clinical microbiology laboratory using MALDI-TOF MS. J Thorac Dis 2014;6:524−33.

[30] Chen JHK, Yam W-C, Ngan AHY, Fung AMY, Woo WL, Yan MK, et al. Advantages of using matrix-assisted laser desorption ionization-time of flight mass

spectrometry as a rapid diagnostic tool for identification of yeasts and mycobacteria in the clinical microbiological laboratory. J Clin Microbiol 2013;51:3981—7.

[31] Westblade LF, Jennemann R, Branda JA, Bythrow M, Ferraro MJ, Garner OB, et al. Multicenter study evaluating the VITEK MS system for identification of medically important yeasts. J Clin Microbiol 2013;51:2267—72.

[32] Ranque S, Normand AC, Cassagne C, Murat JB, Bourgeois N, Dalle F, et al. MALDI-TOF mass spectrometry identification of filamentous fungi in the clinical laboratory. Mycoses 2014;57:135—40.

[33] Riat A, Hinrikson H, Barras V, Fernandez J, Schrenzel J. Confident identification of filamentous fungi by matrix-assisted laser desorption/ionization time-of-flight mass spectrometry without subculture-based sample preparation. Int J Infect Dis 2015;35:43—5.

[34] McMullen AR, Wallace MA, Pincus DH, Wilkey K, Burnham CA. Evaluation of the VITEK MS matrix-assisted laser desorption ionization—time of flight mass spectrometry system for identification of clinically relevant filamentous fungi. J Clin Microbiol 2016;54:2068—73.

CHAPTER FOUR

Work Procedures in MALDI-TOF Technology

Yannick Hoyos Mallecot[1,2]

[1]CHU Bicêtre Bacteriology-Hygiene Unit Assistance Publique Hôpitaux de Paris (AP-PH),
Le Kremlin-Bicêtre, France
[2]Associated French National Reference Center for Antibiotic Resistance:
Carbapenemase-Producing Enterobacteriaceae, Le Kremlin-Bicêtre, France

4.1 INTRODUCTION

Matrix-assisted laser desorption/ionization—time-of-flight (MALDI-TOF) mass spectrometry should be considered as a clinical microbiology diagnostic device. According to this definition, these devices need to be approved by official entities to ensure the safety and efficacy [1]. Therefore, this chapter describes the procedures for the use of the two systems (VITEK MS and MALDI Biotyper CA) that are currently approved by the Food and Drug Administration (FDA) and have the Conformité Européene (CE) marking [2—4].

4.2 REAGENTS AND EQUIPMENT

4.2.1 VITEK MS

4.2.1.1 Reagents
1. Calibrator: *Escherichia coli* ATCC 8739
2. Matrix: α-cyano-4-hydroxycinnamic acid (VITEK MS-CHCA)
3. Extraction: Formic acid (VITEK MS-FA)

4.2.1.2 Supplies
1. Target slides: Single-use, disposable slides consisting of three acquisition groups, each with 16 sample spots (VITEK MS target slides); each group includes one calibration spot.

2. Wooden application sticks, pipet tips, or plastic inoculation loop to apply material
3. 50−1000 µL pipet tips and a suitable pipet
4. 2−200 µL pipet tips and a suitable pipet
5. 0.5−10 µL pipet tips and a suitable pipet
6. Eppendorf tubes

4.2.1.3 Equipment

1. Instrument: VITEK MS (original equipment manufacturer-labeled Shimadzu AXIMA Assurance mass spectrometer); floor model with a Class 1, 337-nm fixed focus, nitrogen laser.
2. Components:
   a. VITEK MS Prep station: Used to prepare the VITEK MS target slides. Simple scanning integrates the ID at the bench with complete barcoded traceability. Easy to use software keeps track of which spots the user has prepared.
   b. VITEK MS Acquisition Station: Directly connected to the VITEK MS instrument. It displays the instrument status and acquires the sample spectra from the instrument; the signal is recorded as a spectrum of intensity versus mass (in Daltons [Da]). The spectra are then sent to MYLA from the acquisition station for analysis.
   c. MYLA Server: Contains the MYLA software and is a middleware solution. For VITEK MS, MYLA is connected to both the VITEK MS Prep Station and VITEK MS Acquisition Station providing complete integration and traceability of all individual patient samples and their results. Connectivity to the Laboratory Information System and other instruments in the laboratory allows all laboratory data to be grouped together and viewed in one place.
3. Database: VITEK MS Knowledge Base; current version (V3.0) of the reference database contains bacteria, fungi, mycobacteria, nocardia, and molds. Each species or species group is represented by an average of >14 isolates/species and average of 36 spectra/species. Raw spectra were obtained from reference isolates grown on multiple media types, under several growth conditions, and from different sample origins.

4.2.2 MALDI Biotyper system

4.2.2.1 Reagents

1. Calibrator: Bacterial Test Standard (BTS). The BTS consists of a manufactured extract of *E. coli* DH5 alpha, which has a characteristic

peptide and protein profile mass spectrum. The extract is spiked with two additional proteins that extend the upper boundary of the mass range covered by BTS. The overall mass range covered by BTS is 3.6 to 17 kDa

2. Matrix: α–cyano-4-hydroxycinnamic acid (HCCA portioned)
3. Acetonitrile (ACN)
4. Deionized water
5. Trifluoroacetic acid (TFA)
6. Absolute ethanol (EtOH)
7. Formic acid (70%)

4.2.2.2 Supplies

1. Target slides: Reusable steel plates (MBT Biotarget 96), which have been developed for the preparation and identification of, test organisms using the MALDI Biotyper System. The target allows the identification of up to 96 test organisms per run and is designed to minimize the risk of cross-contamination. Target slides must be cleaned following each analysis run.
2. Wooden application sticks, pipet tips, or plastic inoculation loop to apply material
3. 50−1000 μL pipet tips and a suitable pipet
4. 2−200 μL pipet tips and a suitable pipet
5. 0.5−10 μL pipet tips and a suitable pipet
6. Eppendorf tubes

4.2.2.3 Equipment

1. Instrument: Microflex LT/SH mass spectrometer; the desktop model, with closed safety covers is a Class 1 Laser product. With the safety cover opened it becomes a Class 4 Laser product. The laser is a 337 nm fixed focus, nitrogen laser.
2. Components:
   a. MBT System client: A software that displays the user interface and guides the user through the system workflow. It also interfaces with the flexControl software for automated acquisition of mass spectra using the Microflex LT/SH instrument.
   b. MBT System Server: This server communicates with the MBT System client and the MBT-DB server. It performs preprocessing on acquired spectra, and matches peaks lists against the Main

Spectrum (reference pattern, (MSP)) for matching and calculates the score value (log (score)).

c. MBT System DB Server: Made to store all information for the MBT System. The MBT-DB maintains spectra data, project data, method data, user management data, reference patterns, and other peak lists plus additional maintenance data.

d. FlexControl Software Package, including GTPS firmware and flexControl acquisition software.

e. GTPS firmware: Communicates with the flexControl PC software, controls and monitors the vacuum, moves the sample carrier and performs the docking of the target plate, controls and monitors high voltages in the ion source, generates trigger signals, and monitors instrument status.

f. FlexControl acquisition software: Communicates with the MBT System client, loads automatic run jobs, communicates with the GTPS firmware, communicates with the laser in the microflex LT/SH instrument, sets the acquisition parameters in the digitizer and reads the acquired data from the digitizer, performs automated data acquisition, evaluates acquired spectra, adjusts the laser power during automatic data acquisition, performs a recalibration of the time-of-flight (TOF) to mass transformation, stores acquired spectra on disk, and performs source cleaning.

3. Database: Current version (V6.0) of the reference database contains bacteria, fungi, mycobacteria, nocardia, and molds. The reference library now includes more than 2,500 species of 433 microorganism genera with a total of containing 6,903 MSPs. The reference library was established using type strains combined with 5 to 38 additional clinical or culture collection strains per species.

4.3 CALIBRATION AND QUALITY CONTROL

4.3.1 VITEK MS

4.3.1.1 Calibration

Escherichia coli ATCC 8739 strain must be incubated for 18−24 hours at 35°C on blood agar. This organism is deposited with VITEK MS-CHCA matrix on positions: xA1, xB1, xC1, of the VITEK MS-DS target slides depending on the number of samples tested (one calibrator per acquisition group of 16 spots). Calibration will occur as the first automatic step in the process of acquisition of the sample, carried out by the acquisition software. The calibration spot must provide an identification 99.9% *E. coli* in the MYLA software. A correct calibration will allow the instrument to move to the top of the acquisition group; failure of calibration will result in an error and the instrument will automatically proceed to the next acquisition group without collecting any sample spectra. After acquiring the spectra for each test sample in the acquisition group, the calibration spot will be automatically checked again.

4.3.1.2 Controls

Two organisms serve as the positive controls (*Enterobacter aerogenes* ATCC 13048 and *Candida glabrata* ATCC MYA-2950). Matrix alone is used for the negative control. If the positive-control organisms are not correctly identified, the test results are considered invalid. If identification is given in the negative control, the control and test isolates must be reinoculated onto a new, target slide and reanalyzed.

4.3.2 MALDI Biotyper CA system

4.3.2.1 Calibration

BTS is used for mass spectrum calibration and optimization as well as a performance control for the identification of microorganisms with the MALDI Biotyper System.

BTS must be solubilized with standard solvent (ACN 50%, water 47.5%, and TFA 2.5%) according to manufacturer instructions. It is strongly recommend to aliquot and frozen the dissolved BTS.

Two BTS control positions on an MBT Biotarget 96 are selected and inoculated with BTS solution. The BTS solution is allowed to dry at room temperature and then overlaid with reconstituted HCCA portioned solution. During system start-up, auto-calibration is performed, if BTS does not meet all required performance specifications, the test run will be invalid.

4.3.2.2 Controls

Appropriate external quality control (QC) should be included depending on test organisms (at least one QC per group if strains of these groups are going to be analyzed).

Any of the following external QC test organisms could be used:

Gram–negative

Enterobacter cloacae ATCC 13047

Klebsiella pneumonia ATCC 13883

Proteus vulgaris ATCC 29905

Pseudomonas aeruginosa ATCC 10145

Escherichia coli ATCC 25922

Gram–positive

Enterococcus faecalis ATCC 17433 T

Enterococcus faecium ATCC 19434 T

Staphylococcus aureus ATCC 12600 T

Haemophilus

Haemophilus influenzae ATCC 49766

If the QC organism is not correctly identified or yields a log (score) <2.00, the run is considered invalid. All isolates and QC organisms have to be respotted using a new target plate and testing repeated.

4.4 PREANALYTICAL PROCESSING METHODS FOR PROTEIN EXTRACTION

4.4.1 Specimen collection

Cultivation conditions have little effect on the identification [5]. Media such as Columbia Blood Agar, Chocolate Agar, or others can be used regardless of different growth phases or temperatures. These different cultivation conditions produce only very small variations in observed peaks. Nearly all peaks are reproducible. The recommended culture incubation time for testing bacteria and yeast must be between 24 and 72 hours. However, studies suggest that microbial identification with matrix-assisted laser desorption/ionization (MALDI)−TOF mass spectrometry (MS) is independent of culture conditions. In general a portion of a microbial colony from an agar plate is applied to a spot on MALDI-TOF MS target slide, then a matrix solution is applied to the spot. Finally the target slide is dried and then loaded into the MALDI-TOF MS instrument. Feasibility studies suggest that most isolates are any more viable after the addition of matrix; however, viability of all organisms in all growth conditions has not been tested yet. Therefore, all specimens must be considered infectious and handled appropriately and universal precautions for the management of organisms should be respected during the whole procedure.

4.4.2 MALDI Biotyper clinical application system extraction protocol

According to the Food and Drug Administration (FDA)/Conformité Européene (CE) approved protocol, organisms (bacteria and yeast) could be tested by using a direct transfer (DT) technique. If the initial analysis results in a low score, organisms may be processed by using either an extended direct transfer (EDT) or a formic acid extraction (EX) procedure.

1. DT method
   a. Smear biological material (single colony) as a thin film directly onto a spot on a MALDI target plate.
   b. Overlay the material with 1 μL of HCCA solution within 1 hour and allow drying at room temperature.
2. EDT Method
   a. Smear biological material (single colony) as a thin film directly onto a spot on a MALDI target plate.
   b. Overlay the material with 1 μL of 70% formic acid and allow drying at room temperature.
   c. Overlay the material with 1 μL of HCCA solution within 1 hour and allow drying at room temperature.
3. EX Method
   a. Pipette 300 μL deionized water into an Eppendorf tube.
   b. Place biological material into the tube. Mix thoroughly. Pipetting or vortexing should be appropriate.
   c. Add 900 μL EtHO and mix thoroughly.
   d. Centrifuge at a maximum speed for 2 minutes, decant supernatant, centrifuge again, and remove all the residual EtOH by carefully pipetting it off to waste without disturbing the pellet.
   e. Allow the EtOH-pellet to dry at room temperature for 2−3 minutes.
   f. Add 70% formic acid (1−80 μL depending on the size of the pellet) to the pellet and mix very well by pipetting and/or by vortexing.
   g. Add pure ACN (same volume as formic acid) and mix carefully.

h. Centrifuge for 2 minutes at a maximum speed (13,000−15,000 rpm), such that all material is collected neatly in a pellet.

i. Pipette 1 μL of supernatant onto a MALDI target plate and allow it to dry at room temperature.

j. Overlay the material with 1 μL of HCCA solution within 1 hour and allow it to dry at room temperature

4.4.3 VITEK MS extraction protocol

According to the FDA/CE approved protocol, no extraction is needed for bacteria prior to slide inoculation. For yeast a VITEK MS-FA reagent is used to extract protein before the VITEK MS-CHCA matrix is added to the spot containing the sample.

4.4.4 Special sample preparation methods (research use only)

4.4.4.1 Aerobic actinomycetes

Identification of aerobic actinomycetes, using MALDI-TOF MS is more difficult because of the stronger cell wall that is not correctly disrupted by standard laser incision. Therefore, identification of nocardia and mycobacteria is significantly improved using mechanical and chemical extraction, either alone or in combination. Several protocols have been described in the literature. Here is an example of selected protocols:

4.4.4.1.1 Nocardia [6]

1. Put 300 μL deionized water into a 1.5 mL Eppendorf tube.
2. Transfer bacteria into the tube (avoid collecting medium). Try to get one to three 10-μL inoculation loops of biomass. To get an idea of the amount of biomass: 2 μL water in an Eppendorf tube represents a small pellet, 5 μL of water represents a good-sized pellet.
3. Boil for 30 minutes in a water bath (95°C in a Thermomixer may also be suitable).
4. Centrifuge for 2 minutes at maximum speed (13,000−15,000 rpm), decant supernatant.
5. Add 300 μL deionized water and 900 μL EtOH, mix thoroughly.
6. Centrifuge for 2 minutes at maximum speed (13,000−15,000 rpm), decant supernatant.

7. Repeat centrifugation and resuspension steps two more times.
8. Allow the pellet to dry at room temperature (e.g., 2 minutes should be sufficient).
9. Add pure ACN (in general 10 μL−50 μL, depending on size of the pellet, if unsure use 20 μL).
10. Use a vortex mixer at maximum speed for 1 minute.
11. Add 70% formic acid (same volume as ACN) and mix by vortexing for approximately 5 seconds.
12. Centrifuge at maximum speed for 2 minutes.
13. Place 1 μL of supernatant on a MALDI target plate and allow drying at room temperature.

4.4.4.1.2 Mycobacteria [7]

Samples of unknown mycobacteria must be considered as *Mycobacterium tuberculosis* complex members, which are harmful. These organisms are classified as biosafety Level 3 organisms and therefore, heat inactivation is mandatory for these samples.

However, because of differences in the properties and performance of laboratory equipment, a complete inactivation of mycobacterial cells cannot be guaranteed.

In one study Lotz et al. [7] reported a complete inactivation of *M. tuberculosis* after 10 minutes of incubation in 70% ethanol. Williams et al. [8] tested *M. tuberculosis* viability in 50%, 70%, and 95% ethanol for 1, 6 and 24 hours, respectively. No growth of *M. tuberculosis* was observed after 1 hour of incubation for all three percentages of ethanol tested.

However, Chedore et al. [9] reported viable *M. tuberculosis* cells after fixing for 10 minutes in 70% ethanol. Lind et al. found that a 15-minute exposure of *M. tuberculosis* to 70% ethanol reduced the titter of *M. tuberculosis* more than 1000-fold.

Finally Zwadyk et al. [10] found that heating of samples below 100°C may not consistently kill mycobacteria. Heating at 100°C in a boiling water bath for a minimum of 5 minutes kills mycobacteria.

The following inactivation and extraction procedure is derived from the literature and combines these different inactivation methodologies. However, it is currently not recommended to use MALDI-TOF MS analysis for the identification of suspected *M. tuberculosis* isolates. If used, users should evaluate the inactivation procedure for *M. tuberculosis* complex isolates for every laboratory and the equipment therein.

1. Solid medium samples
   a. Pipette 300 μL deionized water into a 1.5 mL Eppendorf Safe-Lock Tube.
   b. Transfer mycobacteria biomass into the Eppendorf Safe-Lock Tube. Try to transfer one to three 10 μL inoculation loops of biomass. To get an idea of the amount of biomass required: 2 μL water in an Eppendorf Safe-Lock Tube represents a small pellet, 5 μL of water represents a suitably sized pellet.
   c. Perform heat inactivation by boiling for 30 minutes.
2. Liquid medium samples for example, MGIT medium
   a. Collect biomass by aspirating 1.2 mL of liquid medium from the bottom of the MGIT tube and transferring the aspirated medium into an Eppendorf Safe-Lock Tube.
   b. Centrifuge for 2 minutes at maximum speed (≥13,000 rpm) and carefully remove the supernatant using a pipette.
   c. Add 300 μL deionized water to the pellet.
   d. Perform heat inactivation by boiling for 30 minutes.
3. Combined follow-up for solid and liquid medium samples
   a. Once inactivation process has been completed, pipette 900 μL EtOH into the Eppendorf Safe-Lock Tube and mix using a vortex mixer.
   b. Centrifuge for 2 minutes at maximum speed (≥13,000 rpm) and decant supernatant.
   c. Centrifuge again and carefully remove residual liquid using a pipette.
   d. Dry the pellet at room temperature (a few minutes should be sufficient).
   e. Add a small amount of Zirconia/Silica beads to the Eppendorf Safe-Lock Tube.
   f. Depending on the size of the pellet, add 10−50 μL pure ACN to the Eppendorf Safe-Lock Tube (if you are unsure of the suitable volume, use 20 μL).
   g. Mix using a vortex mixer at maximum speed for 1 minute.
   h. Add a volume of 70% formic acid equal to the volume of ACN and mix using a vortex mixer for 5 seconds.
   i. Centrifuge at maximum speed (≥13,000 rpm) for 2 minutes.
   j. Place 1 μL of supernatant on a MALDI target plate and allow drying at room temperature.
   k. Immediately after the sample spot has dried, overlay the spot with 1 μL of HCCA matrix.

4.4.4.2 Filamentous fungi [11]

Unlike for yeasts, MALDI-TOF MS−based identification of filamentous fungi has been limited over the last years, due mainly to the requirement for extended sample preparation to achieve good-quality mass spectra. To date, the optimal procedure for identification of filamentous fungi using MALDI-TOF MS has yet to be determined.

Because fungi used to attach very strongly with agar-media, frequently harvesting without picking agar contaminations is not possible. Therefore, cultivation in liquid medium is beneficial.

One proposed sample preparation method and extraction method is as follows:

1. Excise mycelium (approximately 5 mm in diameter) from solid agar and inoculate tubes containing 8 mL of Sabouraud liquid broth.
2. Incubate in a shaker until enough biological material is observed (24−48 hours).
3. Remove cultivation tubes from the rotator, place on the bench, and wait for 10 minutes.
4. Harvest up to 1.5 mL from the sediment and transfer it to an Eppendorf tube.
5. Centrifuge for 2 minutes at full speed (13,000 rpm), then remove the supernatant.
6. Add 1 mL deionized water to the pellet and vortex for 1 minute.
7. Centrifuge for 2 minutes at full speed (13,000 rpm), then remove the supernatant.
8. Suspend the pellet in 300 µL deionized water, add 900 µL EtOH, and vortex it.
9. Centrifuge for 2 minutes at full speed (13,000 rpm), then remove the supernatant.
10. Depending on the size of the pellet, add 10−50 µL 70% formic acid to the Eppendorf Safe-Lock Tube (if you are unsure of the suitable volume, use 20 µL).
11. Add the same volume of ACN to the tube and mix it carefully.
12. Centrifuge for 2 minutes at full speed (13,000 rpm).
13. Place 1 µL of supernatant on a MALDI target plate and allow drying at room temperature.
14. Immediately after the sample spot has dried, overlay the spot with 1 µL of HCCA matrix.

4.5 ANALYTICAL PROCEDURE

4.5.1 MALDI Biotyper system

4.5.1.1 Inoculation

1. Using a sterile inoculating device, remove a portion of the sample (1 µL of the supernatant if an extraction has been performed or an individual colony if DT).
2. Apply sample as a thin layer in each testing spot of an MBT Biotarget 96. Both insufficient and excessive quantity of colony applied can lead to identification failures.
3. Repeat inoculation process for each control and sample to be tested.
4. Allow target slide to air dry.
5. Apply 1 µL of HCCA portioned (matrix) to each testing spot.
6. Allow spots to air dry at room temperature.
7. Test sample and control/calibrator spots on target plate must be manually identified in software. All other analysis steps are performed automatically by the MBT-CA system.

4.5.1.2 Spectrum acquisition

The MBT system software automatically performs spectra acquisition process. During this process, the control and test samples are exposed to multiple laser shots to acquired protein mass spectral profiles.

4.5.1.3 Identification

The MBT system for identification is based on unique protein patterns of the microorganisms obtained from MS. The test organism's spectrum (a pattern of mass peaks) is compared with a reference spectra library (database). Using biostatistical analysis, a probability ranking of the organism identification is generated.

The whole process is as followed:

1. During the automatic spectra acquisition process, the spectrum of an unknown organism is transformed into a peak list.
2. The peak list is compared with the reference peak list of each organism found in the reference library and a log (score) is generated (the higher log score, the higher degree of similarity to the organism listed on the database).

3. Identification results:
   - Log (score) ≥ 2.000: Organism identification is reported with high confidence (allows for species-level identification).
   - Log (score) between 1.700 and 2.000: Organism identification is reported with low confidence (allows for genus-level identification).
   - Log (score) <1.700: No identification.

4.5.2 VITEK MS

4.5.2.1 Inoculation

1. Using a sterile inoculating device, remove a portion of the microbial colony (bacteria or yeast) from the solid media.
2. Apply sample as a thin layer to a target slide spot.
3. Repeat inoculation process for each control and sample to be tested.
4. Application of matrix:
   a. Bacteria: Apply 1 μL of VITEK MS-CHCA matrix to each control and sample spot.
   b. Yeast: Apply 0.5 μL of VITEK MS-FA to each sample spot and allow it to dry, then add 1 μL of VITEK MS-CHCA matrix to the same spot.
5. Allow spots to air dry at room temperature.
6. Target slide and sample barcodes should be read on the VITEK MS Prep Station to identify inoculated spots. The slide data are then sent to MYLA. All other analysis steps are performed automatically by the VITEK MS.

4.5.2.2 Spectrum acquisition

VITEK MS automatically performs spectra acquisition process. The control and test samples are exposed to multiple laser shots to acquired protein mass spectral profiles that are summed into a single, raw mass spectrum (100 acceptable mass spectral profiles per sample is desired, but a minimum of 30 profiles is required). The spectrum is then processed by baseline correction, denoising, and peak detection to identify well-defined peaks.

4.5.2.3 Identification

The VITEK system for identification uses an "Advanced Spectra Classifiers" to accurately identify microbial isolates. This process produces a confidence value that represents the percent probability that the unknown isolate has been correctly identified.

Once an unknown organism's raw spectrum is acquired by the mass spectrometer and preprocessed, it goes through a mass binning to get an identification.

The whole process is as follows:

1. A weighting algorithm computationally sorts the mass spectral entries for individual isolates into unique bin matrices.
2. Bins are then given weights according to importance; bins with mass spectral data that are highly specific for an individual species are given higher weights, while bins with less specific data are given lower weights.
3. The scores within each individual bin are multiplied by the weighted bin value for each reference species in the Knowledge Base and the sum of the weighted bin scores is calculated.
4. The weighted bin sum is used to determine the confidence values of the unknown isolate relative to each reference species.
5. Identification results:
   a. Single identification: A single identification is displayed as a green light when only one significant organism or organism group is retained with a confidence value ≥ 60.0.
   b. Low discrimination identification: A low-discrimination identification is displayed as a yellow light when more than one significant organism or organism group is retained, but not more than four.
   c. No identification: Nonidentified is displayed as a red light when more than four organisms or organism groups are found (In this case, a list of possible organisms is displayed and the sum of confidence values is less than 100), or as nonidentified U, unclaimed identification when no match is found.

4.6 CLEANING MALDI-TOF MS TARGET PLATES

For the VITEK system, target slides are single-use, disposable slides; therefore, no maintenance is needed. In contrast, the MALDI Biotyper System uses reusable steel plates for measurement and identification organisms. The procedure below must be performed before identification of test organisms using the MALDI Biotyper System.

1. Transfer the target into a suitable container (e.g., a 100 mm glass Petri dish) and pour in enough 70% aqueous ethanol to cover the target.
2. Incubate for 5 minutes at room temperature.
3. Remove the target and rinse it thoroughly under running tap water.
4. Clean the target thoroughly with 70% aqueous ethanol.
5. Rinse the target with tap water and wipe it with a cleaning wipe.
6. Cover the target with a layer of 80% aqueous TFA by adding 100 μL with a pipette and thoroughly wipe all target positions with a cleaning wipe.
7. Rinse the target with high-pressure liquid chromatography—grade water and wipe it dry with a cleaning wipe.
8. Let the target dry completely for at least 15 minutes at room temperature.
9. Store the clean target in the container provided

REFERENCES

[1] Shawar R, Weissfeld AS. FDA regulation of clinical microbiology diagnostic devices. J Clin Microbiol 2011;49(9 Suppl):S80−4.
[2] Bruker Daltonics. K142677 Substantial equivalence determination decision memorandum. Billerica, MA: Bruker Daltonics.
[3] Bruker Daltonics. K130831 Substantial equivalence determination decision summary. Billerica, MA: Bruker Daltonics.
[4] BioMérieux. K124067 Decision summary. Marcy l'etoile, France: Bruker Daltonics.
[5] Valentine N, Wunschel S, Wunschel D, Petersen C, Wahl K. Effect of culture conditions on microorganism identification by matrix-assisted laser desorption ionization mass spectrometry. Appl Environ Microbiol 2005;71(1):58−64.
[6] Verroken A, Janssens M, Berhin C, Bogaerts P, Huang T-D, Wauters G, et al. Evaluation of matrix-assisted laser desorption ionization-time of flight mass spectrometry for identification of nocardia species. J Clin Microbiol 2010;48(11):4015−21.
[7] Lotz A, Ferroni A, Beretti J-L, Dauphin B, Carbonnelle E, Guet-Revillet H, et al. Rapid identification of mycobacterial whole cells in solid and liquid culture media by matrix-assisted laser desorption ionization-time of flight mass spectrometry. J Clin Microbiol 2010;48(12):4481−6.
[8] Williams DL, Gillis TP, Dupree WG. Ethanol fixation of sputum sediments for DNA-based detection of Mycobacterium tuberculosis. J Clin Microbiol 1995;33(6):1558−61.
[9] Chedore P, Th'ng C, Nolan DH, Churchwell GM, Sieffert DE, Hale YM, et al. Method for inactivating and fixing unstained smear preparations of mycobacterium tuberculosis for improved laboratory safety. J Clin Microbiol 2002;40(11):4077−80.
[10] Zwadyk P, Down JA, Myers N, Dey MS. Rendering of mycobacteria safe for molecular diagnostic studies and development of a lysis method for strand displacement amplification and PCR. J Clin Microbiol 1994;32(9):2140−6.
[11] Schulthess B, Ledermann R, Mouttet F, Zbinden A, Bloemberg GV, Böttger EC, et al. Use of the Bruker MALDI Biotyper for identification of molds in the clinical mycology laboratory. J Clin Microbiol 2014;52(8):2797−803.

Indications, Interpretation of Results, Advantages, Disadvantages, and Limitations of MALDI-TOF

Carmen Liébana-Martos
Department of Microbiology, Complejo Hospitalario de Jaén, Jaén, Spain

5.1 INTRODUCTION

The chapter is divided into four sections that address the aspects of indications of the technique in the field of clinical microbiology; the way of interpreting the results obtained to obtain adequate results; a view of the main advantages and disadvantages presented by the technique in the identification of microorganisms; and the limitations that the technique has until now due to both the technique itself and the nature of the samples and that can constitute areas for improvement and development in the use of matrix-assisted laser desorption ionization—time-of-flight (MALDI-TOF) in clinical microbiology. This chapter aims to give an overview of the MALDI-TOF mass spectrometry (MS) technique, assessing both its advantages in the different applications in the field of clinical microbiology and the possible disadvantages and limitations that may arise in its use or interpretation of spectra to be able to evaluate the suitability of its application to obtain adequate results.

5.2 INDICATIONS FOR THE USE OF MALDI-TOF IN CLINICAL MICROBIOLOGY

The sequence of each protein that has been expressed in a microbial cell is encoded in the genome, so the analysis of the proteins contained in living cells can be considered as an indirect genome analysis [1].

The Use of Mass Spectrometry Technology (MALDI-TOF) in Clinical Microbiology.
DOI: https://doi.org/10.1016/B978-0-12-814451-0.00005-8

MALDI-TOF MS can be used for accurate and rapid identification of various microorganisms, such as Gram-positive bacteria, Enterobacteriaceae, nonfermenting bacteria, anaerobes, and even mycobacteria and yeasts [2,3].

MALDI-TOF MS systems are accurate and give reproducible results for the identification of microbes at the species and subspecies levels wellknown human pathogens as *Neisseria, Listeria*, etc., or uncommonly recognized bacterium [4] frequently misidentified by phenotypic tests [5,6].

Most importantly, with MALDI-TOF MS, a first tentative identification result can be reported at a significantly earlier time point, thereby considerably improving sepsis treatment. MALDI-TOF MS-based identification is generally achieved at the species level in a short time. The identification and predecible susceptibility pattern can be reported to the clinician and facilitates the initiation or modification of treatment [3,7]. An earlier microbiological diagnosis with MALDI-TOF MS technology contributes to reduced mortality and hospitalization time with a significant impact on cost savings and public health [8].

Because of the nature of the potential bioterrorism agents, most laboratories are inexperienced in their identification or are forced to rely on multiple phenotypic, immunological, and genotypic identification protocols. MALDI-TOF MS can be useful for the rapid identification and typing of these microorganisms that have significant public health and medical implications [9].

5.3 INTERPRETATION OF RESULTS

The degree of ionization as well as the mass of the proteins determines their individual TOF. Based on this TOF information, a characteristic spectrum is recorded and constitutes a specific sample fingerprint, which is unique for a given species [10].

An organism-specific spectral fingerprint is generated and compared with thousands of spectral profiles in a database to obtain an identification of the organism. The software which compares the spectra generates a numerical value (score value) based on the similarities between the observed and stored data sets. This score value provides information about the validity of the identification. A score value above 2.0 is generally considered to be a valid species level identification. Values between 2.0 and

1.7 represent reliable genus level identifications. Furthermore, the software displays additional results next to the best match for plausibility checks [10,11]. Generally score >2.0 is considered for species-level identification, although lower cutoffs have been used in the literature.

Not all organisms in the databases are represented equally, nor are the spectra present obtained solely from clinical strains. The organism identification by MALDI-TOF MS is critically dependent on the quality and accuracy of the database used. A recent study demonstrated that supplementation of the commercial Biotyper database with additional in-house generated spectra from 229 routine clinical isolates (including Gram-positive and Gram-negative bacteria, yeast, and anaerobes) increased the identification rate from 87.1% to 98% for 498 prospectively analyzed clinical isolates [5].

For species-level identification, the size range generally used is between 2 and 20 kDa as it was found to be very stable and with a strong signal to noise ratio. This size range is dominated by ribosomal proteins which ionize well, provide accurate spectra, and are only minimally influenced by microbial growth conditions. The computer software automatically compares the collected spectra with a reference databank containing a wide variety of medically relevant isolates. The measured spectra are subject to method-inherent noise and, therefore, will never be exactly identical for an individual isolate. It is important to take into account that a complex variety of the biological processes can cause the protein masses to differ from those masses predicted from the genome [12].

The identification is usually carried out by the automatic method. However, a manual method can be used to select those areas of the target plate must be well exposed to the short laser pulses to obtain a more reliable identification. In any case, the result of the identification obtained must be supported by a clinical suspicion and the expert opinion of the microbiologist.

5.4 ADVANTAGES AND DISADVANTAGES

5.4.1 Advantages

MALDI-TOF MS has revolutionized the identification of microbial species in clinical microbiology laboratories. A single MALDI-TOF MS system can be used for Gram–positive bacteria, Gram–negative bacteria, and yeast, which is not the case with biochemical differentiation methods. A few studies indicate that MALDI–TOF MS can even be applied to identify higher eukaryotes such as algae, insects, and nematodes [13].

The key advantages are simplicity, robustness, and accuracy.

Conventional differentiation methods rely on biochemical criteria and require additional pretesting and lengthy incubation procedures. In comparison, MALDI–TOF MS can identify bacteria and yeast within minutes directly from colonies grown on culture plates. Starting from a colony on an agar plate, the identification can be performed without any previous knowledge either of microbiology or of mass spectrometry [13].

With this procedure, the bacterial cells are extracted directly on the sample plate without the need of laborious protein extraction, and individual proteins can be detected directly from mixtures without preceding separation steps.

Typically, isolated colonies of microbes grown on a solid agar medium are prepared by direct deposition of a small amount of cell material onto the target plate for non–acid-fast bacteria. For yeasts and Gram–positive cocci in some cases, on-target extraction with formic acid is applied to improve extraction. For molds and acid-fast bacteria, a multistep extraction procedure in a reaction tube generally improves the spectrum quality and hence identification performance. In the case where a direct deposit does not yield an identification result, an on-target or tube extraction method likely improves the spectrum quality without increasing the risk of obtaining false-positive results [1,13].

Of much higher importance is the reproducibility of the detection of individual proteins or corresponding peaks, respectively, in a particular microbial strain. Those proteins that can be unambiguously identified in intact cell mass spectra are generally structural proteins, that is, proteins that act as structuring elements in the living cell. In consequence, mass fingerprints of microorganisms growing exponentially are remarkably stable and largely independent of factors such as growth media, temperature, and oxygen supply as well as of analytical conditions such as instrumentation, amount of biomass per sample, and applied matrix compound at least for most prokaryotes and yeasts (not for filamentous molds) [1].

Currently used MALDI–TOF MS techniques are nearly independent of culture conditions. Selective media such as MacConkey and XLD Agar can also be used in addition to standard media formulations such as Columbia and Chocolate Agar [4].

A retrospective study of 1116 clinical isolates comparing MALDI-TOF MS with conventional biochemical testing systems showed correct species identification of bacteria in 95.2% of the cases [14].). In a similar prospective study including 1660 bacterial isolates from 109 different species, 84.1% were correctly identified by MALDI-TOF MS at the species level [15].

In general, and for most microbial species, approximately 104−106 cells per sample is the minimum biomass required to yield a spectrum of sufficient quality for identification [1,10].

This methodically simple approach profoundly reduces the cost of consumables and time spent on diagnostics. If only one sample is to be measured, it can be processed in a few minutes. If a target plate containing 96 isolates is used, results can be obtained in about 1 hour [10]. Even in cases in which a second spectrum acquisition may be necessary, for example, after a more elaborate extraction procedure, application of another matrix, or a change of the covered mass range, MALDI-TOF MS remains a rapid and relatively inexpensive tool for potential identification of health care−associated outbreaks caused by various surveillance organisms [16].

The flexibility of the system is another advantage. It provides a high level of customized features, pertinent to their respective fields of interests. The reference spectra database can be amended and edited either by commercially available software updates or by internal laboratory personnel. An open-source platform can be established as a way for users to exchange spectra of isolates to increase their own reference databases. However, the quality of the entries must be controlled to ensure that no wrong data are distributed [10].

Another aspect of MALDI-TOF MS−based identification is the easy and relatively low cost of a single identification result, which permits the selection of multiple if not all colonies from agar plates. This will lead to a widened view of infectious and commensal microorganisms and, possibly, to the recognition of new pathogens or specific disease-related microbial consortia [1].

5.4.2 Disadvantages

For proteome-based identification, the analyzed protein pattern should be stable and influenced by growth conditions only to a limited degree to allow the construction of a database with species-specific protein mass fingerprints. The stability of mass fingerprints is dependent on the selected mass range. This approach, however, has its limitations when multiple proteins match with a measured mass to charge ration (m/z) value [17].

A first obstacle is the comparatively small number of described microbial prokaryotic and fungal species compared with estimates of the number of species existing on earth: in bacteria, for example, some 10,000 validly described species versus some estimated hundreds of millions of species. In clinical microbiology, this ratio is less pronounced due to a relatively high number of species that have been described for the reason that potential pathogens are of highest interest. Nevertheless, even in clinical microbiology, there will be a certain share of isolates that cannot be assigned to a species, simply for the reason that this species has not yet been formally described and there are continuing discussions on the need for improved database quality and the inclusion of additional microbial species.

Fresh as possible (not more than 48 hours) colonies can be used for MALDI-TOF identification because with increasing cultivation time, weaker and less distinguished peaks will appear in the spectra. This effect is probably due to ribosomal protein degradation and leads to less efficient species identification [13].

Optimal sample preparation has not yet been achieved for several applications. Different protocols have been employed, but effects of experimental factors, such as culture condition and sample preparation, on spectrum quality and reproducibility, have to be examined. In some studies broth cultures further improved spectrum quality in terms of increasing the number of peaks. In addition, protein extraction methods increased reproducibility in samples prepared using identical culture conditions. MALDI-TOF imaging data suggested that the improvement in reproducibility may result from a more homogeneous distribution of sample associated with the broth treatment. Broth culture plus protein extraction method treatment also yielded the highest rate of correct classification [18,19].

All extraction procedures ensure sufficient lysis of the cells and release of proteins. Typical sample preparation methods for particular types of samples are outlined. A particular sample preparation protocol is required for highly pathogenic species [1]. For yeasts, and Gram-positive cocci in some cases, on-target extraction with formic acid is applied to improve extraction. For molds and acid-fast bacteria, a multistep extraction procedure in a reaction tube generally improves the spectrum quality and hence identification performance. Common methods involve growth in a liquid medium such as cultivation of *mycoplasma* in a broth from which the bacteria are subsequently concentrated and, if necessary, washed before mass spectrum acquisition and subsequent identification [19,20].

A particular sample preparation protocol is required for highly pathogenic species (e.g., tuberculous mycobacteria and other organisms requiring biological safety level (BSL 3 conditions). Because MALDI-TOF MS systems are not typically found in BSL 3 laboratories, any suspected BSL 3 sample that is to be analyzed must be completely inactivated before being handled at a lower safety level [1].

MALDI-TOF MS fingerprints are clearly different from genomic data. While the latter have an essentially digital nature, MS peaks are waveform data that are more analog in nature. The presence versus absence, m/z, and relative intensity levels are subject to analytical error, biological and technical variation including complex, sometimes low-level protein expression, posttranslational modification and its regulation, and analyte incorporation in matrix crystals. Although a high intraspecific similarity of mass fingerprints is generally the case for most species, a generalization should only be made with care because for particular species a pronounced variability of mass spectral patterns can be observed [13,21].

Such differentiation results therefore are closely related to the results of the 16 s rDNA sequence database comparisons. Consequently, species which do not differ sufficiently in their ribosomal protein sequences, such as *Shigella* spp. and *Escherichia coli* or *Streptococcus pneumoniae* (pneumococcus) and members of the *Streptococcus oralis/mitis* group, cannot be distinguished by MALDI-TOF MS.

For Gram-negative anaerobic cocci a comparison of mass fingerprints to sequence data revealed that intraspecific variability was found most pronounced in the same species. High intraspecific variability of mass fingerprints can also be observed for certain filamentous fungi.

Examples are the failure to discriminate between *E. coli* and *Shigella* spp. or difficulties to discriminate species of the mitis complex of *Streptococci*. For the *Shigella/E. coli* problem, the reason is simply that, from a phylogenetic point of view, the four *Shigella* spp. are particular types of *E. coli*, most closely related to enteroinvasive *E. coli*. The delineation of *S. pneumoniae* and *S. mitis* is partly hampered by the close relationship of the two species especially for clones of the transition zone between the two species [13].

The various growth forms of molds, such as mycelium and conidia, complicate the analysis due to differences in protein composition [10].

Finally, during routine identification testing, unexpected results are regularly obtained, and the best methods for transmitting these results into clinical care are still evolving. We here discuss the success of MALDI-TOF MS in clinical microbiology and highlight the fields of application that are still amenable to improvement [1].

5.5 LIMITATIONS OF MALDI-TOF TECHNIQUE

5.5.1 Typing

MALDI-TOF has already been thoroughly evaluated for the identification of clinically relevant bacterial species with adequate to excellent results. However, at this point, it is not clear whether the typing of microorganisms by MALDI-TOF MS (MALDI typing) will be as successful as MALDI identification [16,22].

When strains from a small number of clonal complexes, sequence types, or serotypes can be distinguished, this cannot always be extrapolated to a larger number of phylotypes. The same applies to strains from geographically close entities [16].

Peaks that are considered type-specific biomarkers should be recorded consistently with reasonable signal intensities to avoid false interpretation based on analytical variability. Furthermore, proteins corresponding to biomarker peaks may not be expressed under different cultivation conditions. Thus, some potentially specific peaks only expressed in a specific medium cannot be used for the purpose of typing. Consequently, typing schemes should be standardized as much as possible or the robustness needs to be tested prior to a transfer to other systems. This includes analyzing a sufficient number of isolates to ensure the validity of specific biomarker—strain associations. The recording of a peak in a mass spectrum cannot be taken as representing unambiguous detection of a particular protein without further analysis [16].

More closely related species, however, show some similarities in mass fingerprints as in the upper three spectra. It needs to be underlined that peaks with similar m/z values in mass spectra not necessarily represent the same protein; therefore, a single peak is not useful for a characterization in most cases. Second, mass fingerprints of multiple representatives of a single species need to be similar to allow to establish a species-specific mass fingerprint consisting of multiple, consistently recorded peaks. For a set of mass fingerprints of multiple strains, a consensus spectrum can be computed in analogy to a consensus sequence of a housekeeping gene, which can be used as a species-specific mass fingerprint [16,17].

Another reason for a delay in the use of MALDI typing likely relates to a lack of guidelines for data interpretation [16].

Some specific groups of organisms such as *E. coli* and *Shigella* sp. or *S. pneumoniae* have proven to be potentially difficult to identify. These difficulties have probably slowed their potential evaluation by MALDI typing. However, primary identification of *S. pneumoniae* using specific instruments and recent success with clone-specific MALDI typing have demonstrated the potential of MALDI typing for species that may be difficult to identify with this technology.

Although recent studies on *Salmonella* spp., *Francisella tularensis*, *Bacteroides fragilis*, and *Streptococcus agalactiae* demonstrated the potential of MALDI-TOF MS as a tool to differentiate bacteria on a subspecies level [23]. Other examples for subspecies typing include the accurate identification of genomic species from the *Acinetobacter baumannii* group and the rapid subtyping of *Yersinia enterocolitica* isolates Espinali). The key remaining question is whether current discrepancies between results obtained through conventional methods and the outcomes of MALDI-TOF MS can be better explained when developing guidelines for new diagnostic and clinical protocols [22–26].

5.5.2 Resistance and virulence factors

Acquisition of virulence or resistance factors which are not directly detectable because of their high molecular mass (m/z of 20 kDa) may not be reflected by changes in whole-cell mass spectra.

New applications such as in the field of antimicrobial susceptibility testing have been proposed but not yet translated to the level of ease and reproducibility that one should expect in routine diagnostic systems.

Until now, MALDI-TOF MS—based methods for detection of sensitivity to antimicrobials are limited to some antibiotics. These methods has been used to determine β-lactamase or carbapenemase activity based on hydrolysis of the antimicrobial agent and require more or less prolonged periods of incubation in the presence of them, so they are not useful in the laboratory routine and need a highly qualified personnel for the interpretation of the results [6,27−29].

5.5.3 Direct sample test

To date, in nearly all cases a preculture is required for successful analysis of most patient samples.

At present, MALDI-TOF MS cannot yet be performed directly on clinical samples in most cases, because the relatively low number of microorganisms presents in the sample does not allow for accurate spectra acquisition. However, after enrichment during a liquid culture phase, identification becomes possible. For positive blood cultures, the microbial biomass is generally sufficient but needs to be concentrated and purified prior to mass spectral analysis [30−34].

As an alternative to these concentration protocols, a so-called "short incubation" method has been applied with success by spreading an aliquot of a positive blood culture on an agar plate and incubating for 4−6 hours followed by sample preparation [1].

Although this method extends the time of obtaining a result for individual samples, the overall efficiency is similar to that of more direct methods, mainly because the hands-on time is short. Nevertheless, these modifications do not abrogate the inherent limitations of the system, such as difficulties distinguishing between *S. pneumoniae* and *S. mitis/ Streptococcus oralis* [1,4,13].

REFERENCES

[1] Van Belkum A, Welker M, Pincus D, Charrier J-P, Girard V. Matrix-assisted laser desorption ionization time-of- flight mass spectrometry in clinical microbiology: what are the current issues? Ann Lab Med 2017;37:475−83.
[2] Sloan A, Wang G, Cheng K. Traditional approaches versus mass spectrometry in bacterial identification and typing. Clin Chim Acta 2017;473:180−5 Elsevier.
[3] van Belkum A, Chatellier S, Girard V, Pincus D, Deol P, Dunne WM. Progress in proteomics for clinical microbiology: MALDI-TOF MS for microbial species identification and more. Expert Rev Proteomics 2015;12:595−605 Informa Healthcare.
[4] TeKippe EM, Shuey S, Winkler DW, Butler MA, Burnham CAD. Optimizing identification of clinically relevant gram-positive organisms by use of the Bruker Biotyper matrix-assisted laser desorption ionization-time of flight mass spectrometry system. J Clin Microbiol 2013;51:1421−7.
[5] Dingle TC, Butler-wu SM. MALDI-TOF mass spectrometry for microorganism identification. Clin Lab Med 2013;33:589−609 Elsevier.
[6] Oviaño M, Sparbier K, Barba MJ, Kostrzewa M, Bou G. Universal protocol for the rapid automated detection of carbapenem-resistant Gram-negative bacilli directly from blood cultures by matrix-assisted laser desorption/ionisation time-of-flight mass spectrometry (MALDI-TOF/MS). Int J Antimicrob Agents 2016;48:655−60 Elsevier.

[7] McElvania TeKippe E. The added cost of rapid diagnostic testing and active antimicrobial stewardship: is it worth it? J Clin Microbiol 2017;55(1):20−3.

[8] Angeletti S. Matrix assisted laser desorption time of flight mass spectrometry (MALDI-TOF MS) in clinical microbiology. J Microbiol Methods. 2017;138:20−9 Elsevier.

[9] Rudrik JT, Soehnlen MK, Perry MJ, Sullivan M, Reiter-Kintz W, Lee PA, et al. Safety and accuracy of matrix-assisted laser desorption ionization−time of 1 flight mass spectrometry (MALDI-TOF MS) to identify highly pathogenic organisms. J Clin Microbiol 2017;55:3513−29.

[10] Wieser A, Schneider L, Jung J, Schubert S. MALDI-TOF MS in microbiological diagnostics-identification of microorganisms and beyond (mini review). Appl Microbiol Biotechnol 2012;93(3):965−74.

[11] Rodríguez-Sánchez B, Alcalá L, Marín M, Ruiz A, Alonso E, Bouza E. Evaluation of MALDI-TOF MS (matrix-assisted laser desorption-ionization time-of-flight mass spectrometry) for routine identification of anaerobic bacteria. Anaerobe 2016;42:101−7.

[12] Lay JO. MALDI-TOF mass spectrometry of bacteria. Mass Spectrom Rev 2001;20 (4):172−94 Wiley Subscription Services, Inc., AWiley Company.

[13] Welker M. Proteomics for routine identification of microorganisms. Proteomics 2011;11:3143−53 WILEY-VCH Verlag.

[14] Eigner U, Holfelder M, Oberdorfer K, Betz-Wild U, Bertsch D, Fahr A-M. Performance of a matrix-assisted laser desorption ionization-time-of-flight mass spectrometry system for the identification of bacterial isolates in the clinical routine laboratory. Clin Lab 2009;55:289−96.

[15] Seibold E, Maier T, Kostrzewa M, Zeman E, Splettstoesser W. Identification of Francisella tularensis by whole-cell matrix-assisted laser desorption ionization−time of flight mass spectrometry: fast, reliable, robust, and cost-effective differentiation on species and subspecies levels. J Clin Microbiol 2010;48:1061−9.

[16] Spinali S, van Belkum A, Goering RV, Girard V, Welker M, Van Nuenen M, et al. Microbial typing by matrix-assisted laser desorption ionization-time of flight mass spectrometry: do we need guidance for data interpretation? J Clin Microbiol 53 2015;760−5 American Society for Microbiology (ASM).

[17] DeMarco ML, Ford BA. Beyond identification: emerging and future uses for MALDI-TOF mass spectrometry in the clinical microbiology laboratory. Clin Lab Med 2013;33(3):611−28.

[18] Goldstein JE, Zhang L, Borror CM, Rago JV, Sandrin TR. Culture conditions and sample preparation methods affect spectrum quality and reproducibility during profiling of Staphylococcus aureus with matrix-assisted laser desorption/ionization time-of-flight mass spectrometry. Lett Appl Microbiol 2013;57:144−50.

[19] Lee H-S, Shin JH, Choi MJ, Won EJ, Kee SJ, Kim SH, et al. Comparison of the Bruker Biotyper and VITEK MS matrix-assisted laser desorption/ionization time-of-flight mass spectrometry systems using a formic acid extraction method to identify common and uncommon yeast isolates. Ann Lab Med 2017;37(3):223−30.

[20] Biswas S, Rolain J-M. Use of MALDI-TOF mass spectrometry for identification of bacteria that are difficult to culture. J Microbiol Methods 2013;92:14−24.

[21] Giebel R, Worden C, Rust SM, Kleinheinz GT, Robbins M, Sandrin TR. Microbial fingerprinting using matrix-assisted laser desorption ionization time-of-flight mass spectrometry (MALDI-TOF MS) applications and challenges. Adv Appl Microbiol 2010;71:149−84.

[22] Sauget M, Valot B, Bertrand X, Hocquet D. Can MALDI-TOF mass spectrometry reasonably type bacteria? Trends Microbiol 2017;25(6):447−55.

[23] Nagy E, Becker S, Kostrzewa M, Barta N, Urbán E. The value of MALDI-TOF MS for the identification of clinically relevant anaerobic bacteria in routine laboratories. J Med Microbiol. 2012;61(Pt 10):1393−400. Available from: https://doi.org/10.1099/jmm.0.043927-0 Epub 2012 Jun 14.

[24] Stephan R, Cernela N, Ziegler D, Pflüger V, Tonolla M, Ravasi D, et al. Rapid specific identification and subtyping of Yersinia enterocolitica by MALDI-TOF massspectrometry. J Microbiol Methods 2011;87(2):150−3. Available from: https://doi.org/10.1016/j.mimet.2011.08.016 Epub 2011 Sep 3.

[25] Murray PR. Matrix-assisted laser desorption ionization time-of-flight mass spectrometry: usefulness for taxonomy and epidemiology. Clin Microbiol Infect 2010;16:1626−30 Elsevier.

[26] Seng P, Drancourt M, Gouriet F, La Scola B, Fournier P, Rolain JM, et al. Ongoing revolution in bacteriology: routine identification of bacteria by matrix-assisted laser desorption ionization time-of-flight mass spectrometry. Clin Infect Dis 2009;49:543−51 Oxford University Press.

[27] Idelevich EA, Sparbier K, Kostrzewa M, Becker K. Rapid detection of antibiotic resistance by MALDI-TOF mass spectrometry using a novel direct-on-target microdroplet growth assay. Clin Microbiol Infect 2017; pii: S1198-743X(17)30578-5.

[28] Knox J, Jadhav S, Sevior D, Agyekum A, Whipp M, Waring L, et al. Phenotypic detection of carbapenemase-producing Enterobacteriaceae by use of matrix-assisted laser desorption ionization-time of flight mass spectrometry and the Carba NP test. J Clin Microbiol 2014;52:4075−7 American Society for Microbiology (ASM).

[29] Lasserre C, De Saint Martin L, Cuzon G, Bogaerts P, Lamar E, Glupczynski Y, et al. Efficient detection of carbapenemase activity in enterobacteriaceae by matrix-assisted laser desorption ionization-time of flight mass spectrometry in less than 30 minutes. J Clin Microbiol 2015;53:2163−71 American Society for Microbiology (ASM).

[30] Ashizawa K, Murata S, Terada T, Ito D, Bunya M, Watanabe K, et al. Applications of copolymer for rapid identification of bacteria in blood culture broths using matrix-assisted laser desorption ionization time-of-flight mass spectrometry. J Microbiol Methods 2017;139:54−60 Elsevier.

[31] Scott JS, Sterling SA, To H, Seals SR, Jones AE. Diagnostic performance of matrix-assisted laser desorption ionisation time-of-flight mass spectrometry in blood bacterial infections: a systematic review and meta-analysis. Infect Dis (Auckl) 2016;48:530−6 Taylor & Francis.

[32] Tan KE, Ellis BC, Lee R, Stamper PD, Zhang SX, Carroll KC. Prospective evaluation of a matrix-assisted laser desorption ionization-time of flight mass spectrometry system in a hospital clinical microbiology laboratory for identification of bacteria and yeasts: a bench-by-bench study for assessing the impact on time to identification and cost-effectiveness. J Clin Microbiol 2012;50:3301−8 American Society for Microbiology (ASM).

[33] Tanigawa K, Kawabata H, Watanabe K. Identification and typing of Lactococcus lactis by matrix-assisted laser desorption ionization−time of flight mass spectrometry. Appl Environ Microbiol 2010;76:4055−62.

[34] van Veen SQ, Claas ECJ, Kuijper EJ. High-throughput identification of bacteria and yeast by matrix-assisted laser desorption ionization-time of flight mass spectrometry in conventional medical microbiology laboratories. J Clin Microbiol 2010;48:900−7 American Society for Microbiology (ASM).

CHAPTER SIX

Quality Control in MALDI-TOF MS Techniques

Enrique Pérez-Navarro[1] and Juan Carlos Rodríguez Díaz[2]

[1]CAI Técnicas Biológicas, Unidad de proteómica, Departamento Microbiología, Universidad Complutense, Madrid, Spain
[2]S. Microbiología, Hospital General Universitario de Alicante-ISABIAL: Universidad Miguel Hernández, Alicante, Spain

6.1 GENERAL DESCRIPTION OF THE OPERATION OF MATRIX-ASSISTED LASER DESORPTION/IONIZATION—TIME-OF-FLIGHT

The introduction of mass spectrometry (MS) in clinical microbiology laboratories has led to a technological revolution of the first magnitude that has substantially reduced the response time of our tests, adding reliability to our reports in many cases. This has resulted in our laboratories having improved in efficiency and resulting in improvement in the diagnosis, treatment, and prevention of infectious diseases.

MS is the analytic technique used to analyze the mass to charge ratio (m/z) of various compounds. During the 1980s, the discoveries of various ionization techniques and detection systems [1–3] provided the technical basis that allowed mass spectrometric detection and analysis of molecules [4].

Such a device in which the general principle of MS is to produce, separate, and detect ions consists of an ionization source where a beam of gaseous ions is formed from a sample, a mass analyzer that allows to separate the ions using their m/z, and a detector to deliver a mass spectrum. The latter shows which ions in which relative quantities are formed.

The ionization of the analyte molecule is performed by the reception or loss of an electron. Traditionally, different vaporization methods have been used to transfer molecules into the gas phase. The most popular methods to produce this ionization were the electro impact and chemical ionization methods. However, these methods had some drawbacks due to the production of significant decompositions and fragmentations of the molecules [5].

The analysis of proteins and oligonucleotides became feasible with the development of "soft" ionization techniques such as electrospray ionization mass spectrometry (ESI MS), where the solved sample is sprayed into an electrical field, and matrix-assisted laser desorption/ionization (MALDI) MS [6].

The industry currently uses several types of analyzers combined with suitable detectors to separate and indicate the analyte ions depending on the ionization method (e.g., quadrupole, ion cyclotron resonance cell (ICR), etc.).

Introducing and establishing ionization techniques such as ESI and MALDI MS became one of the most important analysis methods with respect to peptide and protein analyses at the end of the 1980s [6]. These techniques allow determining very precisely the molecule masses of large labile molecules, such as protein molecules.

In MALDI MS, the ionization of the analyte molecules is performed based on the ionization—desorption of the co-crystallized analytes with an organic light-absorbing matrix (α-cyano-4-hydroxy cinnamic acid or sinapinic acid). The laser energy causes structural decomposition of the irradiated crystal and generates a particle cloud.

The ions are accelerated through the electric field in a vacuum flight tube. This field is generated by an electrode, which is mounted some millimeters apart opposite to the sample position. Depending on the polarity, positively or negatively charged ions are accelerated from the sample surface towards the analyzer.

Analyzers associated with MALDI experiments are time-of-flight (TOF) instruments where the mass determination in the high vacuum area is performed by a very precise measurement of the period of time after the acceleration process of the ions in the source and their impact on the detector.

An electrostatic field accelerates the ions formed during a short laser pulse inside the source to a kinetic energy. After leaving the source, the ions pass a field-free drift region in which they are separated due to their m/z ratio. This takes place because at some fixed kinetic energy, ions with different m/z values are accelerated in the ion source to different velocities. Knowing the acceleration voltage and the length of the drift region, the m/z ratio can be determined by measuring the flight time, which is low for small molecules than for bigger ones (their initial energies are identical).

Because predominantly single-charged non-fragmented ions are generated, parent ion masses can easily be determined from the resulting spectrum without the need for complex data processing. The masses are accessible as numerical data for direct processing and subsequent analysis.

MALDI-TOF MS measured experiments are in the range of several microseconds. For that reason, this analytical method has successfully been used in research to determine the mass of proteins and peptides in addition to identifying glycoproteins, oligonucleotides, carbohydrates, and small molecules [3], being one of the most important techniques for the identification of microorganisms.

When MALDI-TOF MS systems have been applied to the microbiological diagnosis in clinical microbiology laboratories, it has been reported that it is capable of identifying more than 90% of the bacteria and fungi that are isolated in the clinical samples and have a clinical significance in human pathologies [7]. The standardization of laboratory protocols for the preparation of samples, the improvement of the analysis software and the high-quality databases of microbial reference mass spectra that are used, it has allowed a significant improvement in the identification of microorganisms, making this technology has often displaced classical phenotype procedures based on the biochemical analysis of microorganisms [8].

In clinical microbiology, the analysis of the ribosomal protein patterns has been used as an identification method since these proteins present variability in the molecular weight depending on the species [9], and in most laboratories, the MALDI-TOF MS system is used. The MALDI-TOF is sold by Bruker Daltonik GmbH and VITEK MS (BioMérieux) companies. Both companies have developed specific databases of human pathogens that not only allow the identification of bacteria but also yeasts, filamentous fungi, and mycobacteria, most of them usually associated with human pathology [10−12].

The identification of the majority of pathogens is already done in many laboratories, but clinical microbiology has other needs that can also be solved using this methodology; although they are not yet fully validated for use in routine clinical practice, the most important ones are:

1. Development of a user-friendly software able to manage huge data volumes quickly and to provide efficient assistance for the interpretation of data. In this important topic, the Computis European project is working to deal it with several developed complementary software tools to process MS imaging data. Data Cube Explorer provides a simple spatial and spectral exploration for MALDI-TOF and

TOF—secondary-ion mass spectrometry (TOF-SIMS) data. SpectViewer offers visualization functions, assistance in the interpretation of data, classification functionalities, peak list extraction to interrogate the biological database and image overlay, and it can process data issued from MALDI-TOF, TOF-SIMS, and desorption electrospray ionization (DESI) equipment [13].

2. The development of bacterial quantification: Quantitative MALDI-TOF MS using an internal standard facilitates the measurement of the quantity of peptides and small proteins within a spectrum. These quantities correlate to the number of microorganisms and therefore to the growth of a microorganism. The main application of this methodology is the development of a system that makes the comparison of the growth in the presence or absence of an antibiotic that allows for analysis of the susceptibility behavior of a strain. Different working protocols and modifications to the instrument as well as special software such as MBT STAR-BL (Bruker Daltonik GmbH, Germany) are being developed; parylene-matrix chip is also being incorporated [14—16].

3. Ability to differentiate clones within the same bacterial species: to control hospital outbreaks and improve the control of multiresistant bacteria, with the aim of developing systems that allow the differentiation of two different clones of the same species [17].

4. Detection capacity of a small number of microorganisms: a strategy based on a DNA-mediated signal amplification for ultrasensitive biomolecular detection is being developed through the use of magnetic beads and sandwiched through the simultaneous interaction with gold nanoparticles and barcode DNA The sandwiched complex was collected by convenient magnetic separation and then treated with potassium cyanide to dissolve the gold nanoparticles and consequently release the DNA molecules, which were then magnetically separated and analyzed by using MALDI-TOF MS [18].

6.2 POSSIBLE SOURCES OF SYSTEM ERROR

Although the identification at the species level has a high degree of reliability in bacteria usually isolated in human pathology, the system poses more problems in isolated bacteria infrequently or with special wall structures, such as mycobacteria. It also poses some problems in the identification of fungi, especially filamentous fungi, associated with the availability of spectra in the databases and the difficulty of processing the samples by the composition of the fungal cell [19].

To minimize these limitations, it is necessary to strictly protocolize the protocols to be followed in each center, keeping in mind that the main sources of error are:

1. Mixture of microorganisms: It is basic to ensure that there is only one species of microorganism in the sample that is deposited in the spectrometer. This is the main source of errors in identification. Currently, work is being done on systems that allow the identification of a mixture of two microorganisms, but they are not yet applied in clinical practice as usual [20]; thus, to better assess the TOF MS detection of polymicrobial samples, databases with combinations of microorganisms have been developed using the most frequent combinations in clinical practice [21].

2. Processing of intact cell or protein extraction: A greater identification is obtained after the protein extraction of the microorganisms, but the processing of the whole cell is faster. If it is an easily identifiable organism, the rapid procedure can be tried, but if a good identification is not achieved, protein extraction should be used. In organisms that are difficult to be identified, it is more appropriate to always perform protein extraction since the results can change depending on the technique used [22,23]. Poor sample preparation will degrade sensitivity and will yield low resolution and poor reproducibility. The generation of ions through MALDI depends on the presence of an optimal ratio of matrix substance to the analyte.

3. Culture media: The type of growth medium can affect the correct identification of the microorganism. In general, better results are obtained if the process is carried out from liquid media or agar plates without dyes or additives and 24−48 hours of incubation. It has also been reported that multiple subcultures cause identification problems [22,24,25]. If the sample contains human proteins, for example, a positive blood culture, it is essential to perform a procedure for purification of non-human material [26].

4. Sample on the spectrometer plate: Thick and thin sample extensions have been evaluated with or without the addition of formic acid, and the addition of a uniform layer with a high amount of microorganisms

and formic acid is the procedure with the best results [24,27]. Although the technique is robust, systems are being developed that contribute to improve the automation in the preparation of the stainless steel plate. This greatly increases the homogeneity of the dried material [28,29].

6.3 QUALITY CONTROL SYSTEMS APPLICABLE BY THE TECHNICIANS OF THE COMMERCIAL HOUSES

MS is a complex technique and is very sensitive to minimal structural modifications of the infrastructure that affect the laser or the measurement and detection of the TOF of the protein particles that are analyzed to achieve the characteristic spectrum of each bacterial and fungal species. This mandates that periodic review must be done by employees of the supplier company to be sure about the correct functioning of the equipment.

In addition, from this basic maintenance, it is important to regulate the mass range selector to determine the range of masses that the equipment can measure according to the needs of the user, determining the limit and upper limit fields. In addition, it is necessary to configure the measurement form since the data can be obtained in a linear or logarithmic mode.

Periodically, a calibration process of the mass spectrometer must be performed. During the calibration, particular masses of the acquired mass spectrum are assigned to reference masses of the loaded reference list (depending on the calibration strategy).

The maintenance team of the device must periodically update the available databases with the team and introduce different software that helps to interpret the data; for example, ClinProTools mass spectrum model helps to identify problematic species such as *Haemophilus* spp. [30].

Another example is the inclusion of statistical analysis software, such as the statistical method of partial least squares discriminant analysis (PLS-DA) applicable to MALDI-TOF MS protein fingerprint data of bacteria; a mass spectral library is created that allows differentiating strains of the same genetically related species [31].

6.4 QUALITY CONTROL SYSTEMS APPLICABLE BY SYSTEM USERS

MALDI-TOF MS—based microbial identification is highly reproducible and can tolerate numerous variables, including differences in testing environments, instruments, operators, reagent lots, and sample positioning patterns [32]. However, since the results are applied in clinical practice without confirmation of the identification of the microorganism by any other technique, it is important to include periodic internal and external controls to ensure the proper functioning of the equipment and the training of the technical staff that manage it. Each laboratory must establish the periodicity of the controls and the composition of the controls according to the care work performed.

Of special interest is the correct identification of certain microorganisms that, due to their technical complexity or their clinical importance, require special attention. It stands out in the identification of *Streptococcus pneumoniae* and its differentiation with other closely related species such as *Streptococcus viridans* since *S. pneumoniae* is genetically similar to the mitis group. The new databases of the two most used systems, Biotyper 3.0 (Bruker) and VITEK MS (bioMérieux), using the ClinProTools software allows the correct identification of microorganisms as it is able to detect the presence of three peaks (6949, 9876, and 9975 *m/z*) that differentiate the species [33]. In spite of these improvements, these microorganisms must be included in all the quality controls that are carried out in the device; both external and internal Bruker Bacterial Test Standard (BTS) contains a carefully manufactured extract of *Escherichia coli* DH5 alpha that shows a characteristic peptide and protein profile in MALDI-TOF mass spectra. The extract is spiked with two additional proteins that extend the upper boundary of the mass range covered by BTS. The overall mass range covered by BTS is 3.6—17 kDa; this control must be included every day when starting the work of the team.

6.4.1 Quality of the database

When MS began to be used in the identification of microorganisms in clinical microbiology laboratories, there were problems in the processing of rare bacteria, yeasts, and fungi. In addition to the possible problems associated with the processing of the samples, the databases used to be very limited. Commercial companies have included more spectra per species and more different species, so this problem has gradually been minimized [34].

Some researchers, with very specific needs, have chosen to design their own databases from well-characterized collections of microorganisms [35,36], for example, SpectraBank database (http://www.spectrabank.org), which provides open access MALDI-TOF mass spectra from a variety of microorganisms [35].

Specific databases of certain pathogens have also been developed, such as the VibrioBase database which was generated from 997 largely environmental strains identified by partial sequencing the *rpoB* (RNA polymerase beta-subunit) gene [37].

Currently, there are many public repositories that allow to access with the name of the protein; in the proteomics field, the most important database is UniProt. UniProt is a catalog of information that combines the sequences of proteins with functional and biological annotations associated with these proteins. UniProt is currently maintained by the European Institute of Bioinformatics (EMBL-EBI), the Swiss Bioinformatics Institute (SIB), and the Protein Information Repository (PIR). Within this resource, the database UniProt KB is of special interest for proteomics (Protein Knowledge Base), which in turn is composed of the TrEMBL and SwissProt databases [38].

6.4.2 Controls when designing a database

When you start your database design, the first thing to analyze is the nature of the application you are designing for. There are many developers that by default apply normalization rules without thinking about the nature of the application and then later getting into performance and customization issues.

The information generated in a typical proteomics experiment used should be organized in three different levels: (i) raw data; (ii) processed results, including peptide/protein identification and quantification values; and (iii) the resulting biological conclusions. Technical and/or biological metadata can be provided for each level independently [39].

It should start from a collection of microorganisms perfectly characterized at the phenotypic and genotypic levels and establish the parameters to ensure the reproducibility of the generated spectra: culture media, incubation time, an extraction procedure, etc. In this example, a database with 11,851 spectra (938 fungal species and 246 fungal genera) has been designed. Validation criteria were established using an initial panel of 422 molds, including dermatophytes, previously identified via DNA sequencing (126 species). In this other example, the design of a database for the *Borrelia* genus, all pathogenic species were selected for humans in Europe (*Borrelia burgdorferi* sensu stricto, *Borrelia spielmanii*, *Borrelia garinii*, and *Borrelia afzelii*) and other species that can affect humans in other geographic locations were added; the dendrogram was obtained by analyzing the protein profiles of the different *Borrelia* species. Another example is the design of the *Leishmania* database, but it is not commercially available [40,41].

6.4.3 Utility of the scores in the identification

The precise implementation of database search algorithms varies by program. For example, one algorithm predicts the fragmentation patterns of all peptides in the database that match the m/z of the precursor of a mass spectrum. The correlation of the predicted ions with the ions of the experimental spectrum determines whether the peptide in the database matches the peptide analyzed in the mass spectrometer. This process is repeated for all mass spectra, and the output is a list of identified peptides with scores indicating how well the spectra matched the predicted spectra.

Experimental goals and samples vary widely from project to project; however, there are some general common themes that run together in most proteomics experiments. First and foremost, you can expect identifications from 10s to 1000s of proteins (including potential modifications) per experiment, depending on the sample being analyzed. There are different quality metrics such as the false discovery rate (i.e. false-positive protein IDs) and protein coverage statistics to help ensure confidence in our results [42].

Depending on the type of search engine you use, you will find different definitions. These are some of them:

- Protein Score: The sum of the ion scores of all peptides that were identified.
- PSMs: The number of peptide spectrum matches. The number of PSMs is the total number of identified peptide spectra matched for the protein. The PSM value may be higher than the number of peptides identified for high-scoring proteins because peptides may be identified repeatedly.
- Coverage: The percentage of the protein sequence covered by identified peptides.

- Proteins: The number of identified proteins in a protein group, that is, the number of proteins displayed in the Protein Group Members view.
- Unique Peptides: The number of peptide sequences that are unique to a protein group. These are the peptides that are common to the proteins of a protein group and that do not occur in the proteins of any other group. The number of unique peptides that determine a protein group can be set in Proteome Discoverer.
- Peptides: The total number of distinct peptide sequences identified in the protein group.
- Xcorr: It (cross correlation) is a measure of the goodness of fit of experimental peptide fragments to theoretical spectra created from the sequence b and y ions.
- Delta CN: Delta Correlation is a measure of the specificity of the fit. It describes how much better the primary candidate's sequence fits the experimental data vs. the secondary candidate's sequence. A smaller number implies a better fit.
- Delta M (ppm): Delta Mass is the deviation of the measured mass from the theoretical mass of the peptide, in ppm.
- q Value: q value is the minimal false discovery rate at which the identification is considered correct
- Posterior Error Probability (PEP): The PEP is the probability that the observed PSM is incorrect.

Based on these technical data, the system marketed by Bruker provides a score that is associated with the quality of the identification of the microorganism; it is considered to be accurate species-level (score of ≥ 2.0) and genus-level (score of ≥1.7) identifications [43]. The VITEK MS System does not offer that information and considers the identification as correct or incorrect.

However, this system of scores is probably too demanding for clinical practice to be habituated by problematic organisms; thus, in studies of identification of anaerobic bacteria and *Aspergillus* spp., it was found that a score of ≥ 1700 is the one that offers the best practical results [44].

REFERENCES

[1] Aston F. LXXII. The mass-spectra of chemical elements. (Part 2). Philos Mag Ser 1920;6(40):628–34.
[2] Busch KL, Cooks RG. Mass spectrometry of large, fragile, and involatile molecules. Science 1982;218:247–54.
[3] Cohen LH, Gusev AI. Small molecule analysis by MALDI mass spectrometry. Anal Bioanal Chem 2002;373:571–86.

[4] Hillenkamp F, Karas M, Beavis RC, Chait BT. Matrix-assisted laser desorption/ionization mass spectrometry of biopolymers. Anal Chem 1991;63:1193A−203A.

[5] Mamyrin BA, Karataev VL, Shmikk D, Zagulin VA. The mass-reflectron, a new nonmagnetic time-of-flight mass spectrometer with high resolution. J Exp Theor Phys 1973;37:45−8.

[6] Karas M, Hillenkamp F. Laser desorption ionization of proteins with molecular masses exceeding 10,000 daltons. Anal Chem 1988;60:2299−301.

[7] Sandrin TR, Goldstein JE, Schumaker S. MALDI TOF MS profiling of bacteria at the strain level: a review. Mass Spectrom Rev 2013;32(3):188−217.

[8] Kliem M, Sauer S. The essence on mass spectrometry based microbial diagnostics. Curr Opin Microbiol 2012;15(3):397−402.

[9] Suarez S, Ferroni A, Lotz A, Jolley KA, Guérin P, Leto J, et al. Ribosomal proteins as biomarkers for bacterial identification by mass spectrometry in the clinical microbiology laboratory. J Microbiol Methods 2013;94(3):390−6.

[10] Wilen CB, McMullen AR, Burnham CA. Comparison of sample preparation, instrumentation platforms, and contemporary commercial databases for MALDI-TOF MS identification of clinically relevant mycobacteria. J Clin Microbiol 2015;53 (7):2308−15.

[11] Deak E, Charlton CL, Bobenchik AM, Miller SA, Pollett S, McHardy IH, et al. Comparison of the Vitek MS and Bruker Microflex LT MALDI-TOF MS platforms for routine identification of commonly isolated bacteria and yeast in the clinical microbiology laboratory. Diagn Microbiol Infect Dis 2015;81(1):27−33.

[12] Jamal WY, Ahmad S, Khan ZU, Rotimi VO. Comparative evaluation of two matrix-assisted laser desorption/ionizationtime-of-flight mass spectrometry (MALDI-TOF MS) systems for the identification of clinically significant yeasts. Int J Infect Dis 2014;26:167−70.

[13] Robbe MF, Both JP, Prideaux B, Klinkert I, Picaud V, Schramm T, et al. Software tools of the Computis European project to process mass spectrometry images. Eur J Mass Spectrom 2014;20(5):351−60.

[14] Lange C, Schubert S, Jung J, Kostrzewa M, Sparbier K. Quantitative matrix-assisted laser desorption ionization-time of flight mass spectrometry for rapid resistance detection. J Clin Microbiol 2014;52(12):4155−62.

[15] Ahn SH, Hyeon T, Kim MS, Moon JH. Gain switching for a detection system to accommodate a newly developed MALDI-based quantification method. J Am Soc Mass Spectrom 2017;28(9):1987−90. Available from: https://doi.org/10.1007/ s13361-017-1711-2.

[16] Oviaño M, Gómara M, Barba MJ, Revillo MJ, Barbeyto LP, Bou G. Towards the early detection of β-lactamase-producing Enterobacteriaceae by MALDI-TOF MS analysis. J Antimicrob Chemother 2017;72(8):2259−62.

[17] Lasch P, Fleige C, Stämmler M, Layer F, Nübel U, Witte W, et al. Insufficient discriminatory power of MALDI-TOF mass spectrometry for typing of Enterococcus faecium and Staphylococcus aureus isolates. J Microbiol Methods 2014;100:58−69.

[18] Ahmad R, Jang H, Batule BS, Park HG. Barcode DNA-mediated signal amplifying strategy for ultrasensitive biomolecular detection on matrix-assisted laser desorption ionization time of flight (MALDI-TOF) mass spectrometry. Anal Chem 2017;89 (17):8966−73.

[19] Balážová T, Makovcová J, Šedo O, Slaný M, Faldyna M, Zdráhal Z. The influence of culture conditions on the identification of Mycobacterium species by MALDI-TOF MS profiling. FEMS Microbiol Lett 2014;353(1):77−84.

[20] Mahé P, Arsac M, Chatellier S, Monnin V, Perrot N, Mailler S, et al. Automatic identification of mixed bacterial species fingerprints in a MALDI-TOF mass-spectrum. Bioinformatics 2014;30(9):1280−6.

[21] Hariu M, Watanabe Y, Oikawa N, Seki M. Usefulness of matrix-assisted laser desorption ionization time-of-flight mass spectrometry to identify pathogens, including polymicrobial samples, directly from blood culture broths. Infect Drug Resist 2017;10:115−20.

[22] Goldstein JE, Zhang L, Borror CM, Rago JV, Sandrin TR. Culture conditions and sample preparation methods affect spectrum quality and reproducibility during profiling of Staphylococcus aureus with matrix-assisted laser desorption/ionization time-of-flight mass spectrometry. Lett Appl Microbiol 2013;57(2):144−50.

[23] Paul S, Singh P, Rudramurthy SM, Chakrabarti A, Ghosh AK. Matrix-assisted laser desorption/ionization-time of flight mass spectrometry: protocol standardization and database expansion for rapid identification of clinically important molds. Future Microbiol 2017;12:1457−66. Available from: https://doi.org/10.2217/fmb-2017-0105.

[24] McElvania Tekippe E, Shuey S, Winkler DW, Butler MA, Burnham CA. Optimizing identification of clinically relevant Gram-positive organisms by use of the Bruker Biotyper matrix-assisted laser desorption ionization-time of flight mass spectrometry system. J Clin Microbiol 2013;51(5):1421−7.

[25] Anderson NW, Buchan BW, Riebe KM, Parsons LN, Gnacinski S, Ledeboer NA. Effects of solid-medium type on routine identification of bacterial isolates by use of matrix-assisted laser desorption ionization-time of flight mass spectrometry. J Clin Microbiol 2012;50(3):1008−13.

[26] Rodríguez JC, Bratos MÁ, Merino E, Ezpeleta C. Utilización de MALDI-TOF en el diagnóstico rápido de la sepsis. Enferm Infecc Microbiol Clin 2016;34(Suppl 2):19−25.

[27] Veloo AC, Elgersma PE, Friedrich AW, Nagy E, van Winkelhoff AJ. The influence of incubation time, sample preparation and exposure to oxygen on the quality of the MALDI-TOF MS spectrum of anaerobic bacteria. Clin Microbiol Infect 2014;20 (12):O1091−7.

[28] Mesbah K, Thai R, Bregant S, Malloggi F. DMF-MALDI: droplet based microfluidic combined to MALDI-TOF for focused peptide detection. Sci Rep 2017;7(1):6756.

[29] Chudejova K, Bohac M, Skalova A, Rotova V, Papagiannitsis CC, Hanzlickova J, et al. Validation of a novel automatic deposition of bacteria and yeasts on MALDI target for MALDI-TOF MS-based identification using MALDI Colonyst robot. PLoS One 2017;12(12):e0190038.

[30] Chen JHK, Cheng VCC, Wong CP, Wong SCY, Yam WC, Yuen KY. Rapid differentiation of Haemophilus influenzae and Haemophilus haemolyticus by use of matrix-assisted laser desorption ionization-time of flight mass spectrometry with ClinProTools mass spectrum analysis. J Clin Microbiol 2017;55(9):2679−85.

[31] Sindt NM, Robison F, Brick MA, Schwartz HF, Heuberger AL, Prenni JE. MALDI-TOF-MS with PLS modeling enables strain typing of the bacterial plant pathogen Xanthomonas axonopodis. J Am Soc Mass Spectrom 2017;28(2):413−21. Available from: https://doi.org/10.1007/s13361-017-1839-0.

[32] Westblade LF, Garner OB, MacDonald K, Bradford C, Pincus DH, Mochon AB, et al. Assessment of the reproducibility of MALDI-TOF mass spectrometry for bacterial and yeast identification. J Clin Microbiol 2015;53(7):2349−52.

[33] Ikryannikova LN, Filimonova AV, Malakhova MV, Savinova T, Filimonova O, Ilina EN, et al. Discrimination between Streptococcus pneumoniae and Streptococcus mitis based on sorting of their MALDI mass spectra. Clin Microbiol Infect 2013;19 (11):1066−71.

[34] Panda A, Ghosh AK, Mirdha BR, Xess I, Paul S, Samantaray JC, et al. MALDI-TOF mass spectrometry for rapid identification of clinical fungal isolates based on ribosomal protein biomarkers. J Microbiol Methods 2015;109:93−105.

[35] Normand AC, Cassagne C, Ranque S, L'ollivier C, Fourquet P, Roesems S, et al. Assessment of various parameters to improve MALDI-TOF MS reference spectra libraries constructed for the routine identification of filamentous fungi. BMC Microbiol 2013;13:76.

[36] Böhme K, Fernández-No IC, Barros-Velázquez J, Gallardo JM, Cañas B, Calo-Mata P. SpectraBank: an open access tool for rapid microbial identification by MALDI-TOF MS fingerprinting. Electrophoresis 2012;33(14):2138−42.

[37] Erler R, Wichels A, Heinemeyer EA, Hauk G, Hippelein M, Reyes NT, et al. MS database for fast identification of Vibrio spp. that are potentially pathogenic in humans. Syst Appl Microbiol 2015;38(1):16−25.

[38] Bateman A, et al. UniProt: a hub for protein information. Nucleic Acids Res 2015;43:D204−12.

[39] Olsen JV, Mann M. Effective representation and storage of mass spectrometry-based proteomic data sets for the scientific community. Sci Signal 2011;4:pe7.

[40] Calderaro A, Gorrini C, Piccolo G, Montecchini S, Buttrini M, Rossi S, et al. Identification of Borrelia species after creation of an in-house MALDI-TOF MS database. PLoS One 2014;9(2):e88895.

[41] Lachaud L, Fernández-Arévalo A, Normand AC, Lami P, Nabet C, Donnadieu JL, et al. Identification of Leishmania by matrix-assisted laser desorption ionization-time of flight (MALDI-TOF) mass spectrometry using a free web-based application and a dedicated mass-spectral library. J Clin Microbiol 2017;55(10):2924−33 pii: JCM.00845-17.

[42] Mann M, Wilm M. Error-tolerant identification of peptides in sequence databases by peptide sequence tags. Anal Chem 1994;66:4390−9.

[43] Lau AF, Drake SK, Calhoun LB, Henderson CM, Zelazny AM. Development of a clinically comprehensive database and a simple procedure for identification of molds from solid media by matrix-assisted laser desorption ionization-time of flight mass spectrometry. J Clin Microbiol 2013;51(3):828−34.

[44] Hsu YM, Burnham CA. MALDI-TOF MS identification of anaerobic bacteria: assessment of pre-analytical variables and specimen preparation techniques. Diagn Microbiol Infect Dis 2014;79(2):144−8.

CHAPTER SEVEN

Application of MALDI-TOF for Bacterial Identification

Sachio Tsuchida
Department of Clinical Laboratory, Chiba University Hospital, Chiba, Japan

7.1 INTRODUCTION

Matrix-assisted laser desorption ionization—time of flight mass spectrometry (MALDI-TOF MS) can facilitate the generation of protein fingerprint signatures from whole bacterial cells [1—3]. Bacteria can be identified using various algorithms by comparing these fingerprints with those in a database of reference spectra [4,5]. Lately, MALDI-TOF MS has been increasingly used as a microbial diagnostic method for identifying the species of pathogens [6,7] and has been performed as a routine assay in laboratories or hospitals for the identification of microorganisms. In addition, it has been directed to the cerebrospinal fluid (CSF) for the rapid identification of pathogens in a patient with bacterial meningitis [8]. The results in clinical laboratory suggest that the MALDI-TOF Biotyper system can precisely detect *Nocardia* in a short time in combination with a simple processing method and in-house database [9].

Although the direct identification of bacteria from the blood culture bottle broth is a promising application of MALDI-TOF MS, blood culture bottles comprise nonbacterial proteins that might interfere with the interpretation of bacterial MS profiles [10], thereby necessitating pretreatment to remove various nonbacterial proteins and concentrate microbes.

MALDI-TOF MS analysis of bacteria cultured for a short time on solid-agar plates inoculated from positive blood cultures enables precise identification, without requiring additional hands-on processing time, especially for the identification of bacteria [4,11,12]. Despite proposition of several conventional protocols, not many are used widely in a clinical context because of complicated operation, involving significant hands-on time.

The Use of Mass Spectrometry Technology (MALDI-TOF) in Clinical Microbiology.
DOI: https://doi.org/10.1016/B978-0-12-814451-0.00007-1
101

Apparently, microbiological diagnosis of bacteremia relies on the subculture of positive blood samples in bottles for 18—24 hours before identification of bacteria by biochemical assay. Overall, the process can require 18—48 hours or more [11,13—16]. Of note, rapid isolation of infectious microorganisms can be paramount for the survival of patients. Thus, novel approaches are required for the rapid analysis of bacteria in clinical microbiology laboratories to improve patient care.

The use of protein profiles obtained directly by MALDI-TOF MS from colonies was successfully implemented. Prod'hom et al. demonstrated the critical first step of separating bacteria from cellular components [17]. Bernhardt et al. developed a more rapid method for isolating bacteria from clinical blood samples rather than from culture in bottles and demonstrated that a filtration system is sensitive for the detection of simulated low-grade bacteria [18].

In this chapter, our proposed review is relatively advantageous, with **application of MALDI-TOF for bacterial identification**.

7.2 HISTORICAL BACKGROUND

Recently, an upsurge has been witnessed in the number of publications related to MALDI-TOF MS applications in medical microbiology, including the identification of isolates, specific antibiotic-resistant profile, and typing of isolates. Although bacterial identification based on MS spectra obtained by MALDI-TOF MS was proposed more than 40 years ago (Fig. 7.1) [19], it found the application as a rapid, inexpensive, and accurate method for identifying isolates that belong to certain bacterial phyla only recently (Fig. 7.1).

In 1975, Anhalt et al. were the first to use MS to identify bacteria [20]. However, MALDI-TOF MS devices designed for the use under customary conditions have only recently been commercially available. For the past few years, MALDI-TOF MS has been used to identify various microorganisms, for example, Gram-negative rods (e.g., *Escherichia coli* and other members of the *Enterobacteriaceae* family), Gram-positive cocci (e.g., *Staphylococcus aureus* and *Streptococcus*), and some Gram-positive rods (e.g., *Bacillus cereus*). The first extensive study to assess the ability of MALDI-TOF MS to identify bacterial strains isolated from clinical samples has been published very

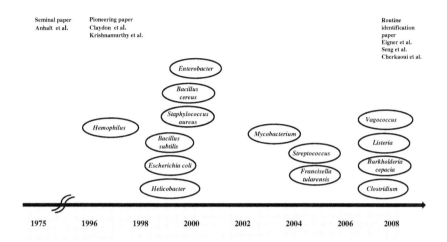

Figure 7.1 Historical background of MALDI-TOF MS applications in medical microbiology.

recently, demonstrating that, at the species level, MALDI-TOF MS precisely identified 84.1% of 1660 strains tested.

In addition, individual strains of a specific organism could also be differentiated easily. Krishnamurthy et al. proposed that some biomarkers correspond to those observed earlier during the MALDI-TOF MS analysis of protein extracts of the same bacteria, which can yield valuable data for rapid classification and detection of microorganisms, representing a substantial breakthrough for rapid screening of environmental as well as biological samples [21]. Claydon et al. reported the application of MALDI-TOF MS for the identification of intact Gram-negative and Gram-positive bacteria obtained directly from culture [22]. In addition, they proposed the analysis of bacteria from a single colony, allowing the screening of mixed cultures, and simple sample preparation with automated analysis that provided spectra within minutes [22]. Their consistent data were obtained from subcultures grown for 3- and 6-day periods from the same cultures 1 day later and from fresh subcultures 2 months later.

About 10 years ago, the proof of concept that MS could identify bacteria was established; however, the pioneering works were published in nonmedical, specialized MS journals. Seng et al. were the first to test an extensive collection of bacteria (1660 bacterial isolates) by MS in a routine laboratory. Of 1660 bacterial isolates analyzed, 95.4% were correctly identified by MALDI-TOF MS; 84.1% were identified at the species level, and 11.3% were identified at the genus level.

Eigner et al. evaluated the performance of the MALDI-TOF MS system for the identification of various clinical isolates in the routine microbiological conditions. Of 108 reference strains, they correctly identified 101 (93.5%) to species level using the MALDI-TOF MS system [23]. The accuracy for the bacterial identification of *Enterobacteriaceae*, nonfermenting Gram-negative rods, *Staphylococci*, *Enterococci*, and *Streptococci* with the MALDI Biotyper system was 95.5%, 79.7%, 99.5%, 100%, and 93.7%, respectively. Furthermore, results were available in 12 minutes for direct smear and in 20 minutes with an extraction method.

Cherkaoui et al. compared two commercially available MALDI-TOF MS devices, MS system (Bruker) and MS system (Shimadzu, Kyoto, Japan) databases, and related analytical tools with common biochemical tests routinely used for the bacterial species identification [24]. While the first MS system (Bruker) provided high-confidence identifications for 680 isolates, of which 674 (99.1%) were correct, the second MS system (Shimadzu) provided high-confidence identifications for 639 isolates, of which 635 (99.4%) were correct.

7.3 MALDI-TOF MS—BASED IDENTIFICATION WORK

Recently, the MALDI Biotyper (Bruker Daltonics GmbH, Leipzig, Germany) was developed as a new system for bacterial identification, in which MALDI-TOF MS is used in combination with database software. Collecting mass spectra of bacteria and comparing these to mass spectra of various bacteria compiled in a database and scoring the match enabled rapid identification of bacteria.

7.4 PERFORMANCE IN ROUTINE CLINICAL MICROBIOLOGY

The diagnosis of bacterial infections and the identification of infecting microorganisms are prerequisites for efficient treatment. In a clinical diagnostic microbiology laboratory, the current method of identifying bacterial isolates is primarily based on phenotypic characteristics, for example, growth pattern in different media, colony morphology, Gram staining, and various biochemical reactions.

The identification of microorganisms in clinical laboratories entails the assignment of a clinical isolate to a genus and, preferably, a species that have been previously defined by a gold standard. To date, various laboratory techniques have been used for the identification of bacteria. Previously, clinical microbiology laboratories relied heavily on conventional methods that often involved culturing, followed by morphological phenotyping and cumbersome biochemical testing.

Although efficient, immunological microbial identification methods have been developed for only a small number of bacterial species. Despite being dominant, molecular-based methods, such as ribosomal RNA sequencing, are complicated and expensive. Conversely, MALDI-TOF MS bacterial identification system is an easy, rapid, accurate, and cost-effective method that has revolutionized bacterial identification in clinical microbiology laboratories. The Sepsityper kit (Bruker Daltonics, Billerica, MA) can facilitate pathogen identification within an hour of positivity when used in conjunction with a MALDI-TOF MS instrument. In addition, Sepsityper can be potentially used in the routine clinical setting. Compared to standard methods of identification, using either Sepsityper or laboratory-developed extraction methods, various studies have reported a reduction in time to presumptive identification of 6–83 hours, depending on the bacterial isolate [25–28].

7.5 IDENTIFICATION OF BACTERIA FROM CULTURE MEDIA

The discussion, thus far, has focused on the identification of bacteria from cultured colonies. Direct analysis of clinical samples without requiring prior culturing might shorten the time required for identifying bacteria and would, therefore, future enhance the utility of the MALDI-TOF MS—based approach. In this approach, a small portion of a cultured colony is directly smeared onto a target plate and overlaid with a matrix solution. Bacterial concentrations of approximately $1 \times 10^4 - 1 \times 10^5$ CFU/mL was set as the limit for MALDI-TOF MS identification [29,30], and a solution of α-cyano-4-hydroxycinnamic acid in a mixture of organic solvents and water is used as the matrix. The spotted mixture is, then, air-dried and inserted into the mass spectrometer for automated measurement. The MS spectra obtained in the analysis of a microorganism are primarily assigned to ribosomal proteins. For the clinical application of MALDI-TOF MS—based bacterial identification, the following two systems (including associated databases) are widely used: the Bruker Biotyper (Bruker Daltonics) and the VITEK MS (bioMérieux). Despite similar analytical principles of the two systems, differences exist in the way databases are constructed and in the algorithms used to identify organisms. The MALDI Biotyper system has already been proven to be comparable, even superior, to classical identification methods in a plethora of peer-reviewed publications. For the VITEK MS system, launched in 2010 and comprising a mass spectrometer from Shimadzu and a database application acquired from AnagnosTec GmbH (Zossen, Germany), a few study reports are already available. Interestingly, a recent study has compared the capability of both systems and established similar results from both platforms [24].

7.6 INAPPROPRIATE SAMPLE PREPARATION

Blood culturing is the leading method for diagnosing bloodstream bacterial infections. However, pretreatment is required for the direct identification of microbes from blood cultures. Conventional protocols, however, are complex and require extensive hands-on time. In addition, such techniques suffer from interference by residual blood debris, including hemoglobin-related proteins. To date, a majority of studies using MALDI-TOF MS for bacterial identification have used the direct-smear method for most isolates and reserved extraction for isolates that were not initially identified; yet, extraction has been proven to enhance identification.

7.7 MS-BASED IDENTIFICATION OF BACTERIA DIRECTLY FROM CLINICAL SAMPLES

7.7.1 Blood

Blood is a complex liquid tissue that comprises various cells and extracellular fluid. In the current research on proteomics, considerable emphasis has been given to the analysis of blood because this source would by far offer the most substantial number of essential biomarker applications. Of note, bacterial and fungal infections of the blood represent a significant public health concern, rendering blood as one of the most critical specimens received by the clinical microbiology laboratory for culture, which is usually the most sensitive method for detecting bacteria. Early positive results from a blood culture can provide valuable diagnostic information on which appropriate antimicrobial therapy can be based. Reportedly, blood culturing is a promising approach for microbial identification with MALDI-TOF MS [31,32]. Usually, patients with blood smears are treated with broad-spectrum antibiotics based on the Gram staining results. Previously, it took an additional day to obtain a definitive identification based on the culture results; however, MALDI-TOF MS facilitates initiating more specific antibiotic treatment shortly after receiving the clinical specimen.

7.7.2 Urine

At present, urine test stripes comprise a central diagnostic instrument with major importance in the first-line diagnostics. These tests detect early symptoms of three groups, diseases of the kidney and the urinary tract, carbohydrate metabolism disorders (diabetes mellitus), and liver diseases and hemolytic disorders. With the advent of the general discussion about the necessity of new biomarkers, biomarker discovery in human urine has become an evolving and potentially valuable field of research in renal function and diseases of the urinary tract. Of note, urinary tract infection is the most common bacterial infection in humans. In a study, of 220 urine samples in which monomicrobial bacterial growth was higher than 10^5 CFU/mL, the organism in 202 samples (91.8%) was identified at the species levels [33]. Veron et al. illustrated a comparison of three of these methods, differential centrifugation, urine filtration, and 5-hour bacterial cultivation on solid culture media, based on their performance of bacterial identification and their potential as a routine tool for microbiology laboratories [34]. A higher percentage of correct MALDI bacterial identification was obtained through filtration (78.9%) and a growth-based method (84.2%) compared to centrifugation method (68.4%). The results of this study demonstrated that short culture is a straightforward and efficient sample preparation method enabling fast and reliable identification of uropathogens by MALDI.

7.7.3 Cerebrospinal fluid

Recently, a number of proteomics analyses in the field of brain research have focused on the CSF as a specimen [35−38]. Usually, the CSF constitutes an adequate specimen for the brain research. As the availability of brain tissue is limited in human research, the CSF is the sample matrix closest to the pathology of the central nervous system (CNS) [39]. Being in direct contact with the CNS, the CSF is a promising source for finding biomarkers for diseases in the CNS. The CSF is a promising target for microbial identification with MALDI-TOF MS. At present, the first step in the detection of bacterial pathogens accountable for meningitis is Gram staining of the CSF, looking for the presence of bacteria. Patients with positive smears are treated with broad-spectrum antibiotics based on the Gram staining results. To date, limited studies have described the use of MALDI-TOF MS for the direct detection of microorganisms causing bacterial meningitis. Besides the clinical examination, the CSF analysis was the crucial neurological diagnostic tool until the introduction of imaging techniques.

7.8 FUTURE DIRECTIONS

7.8.1 Detection of antibiotic resistance

Although organisms' identification can guide antibiotic therapy, several organisms exhibit unpredictable resistance patterns. Thus, the MALDI-TOF MS analyses should be extended to search for antibiotic-resistance determinants. In fact, some studies have reported establishing a correlation between antibiotic resistance and identified bacteria. Although MALDI-TOF MS can be used to detect resistance mechanisms, it is labor-intensive at present. Hence, further research is warranted to establish the application of MALDI-TOF MS to detect antibiotic resistance on a routine basis.

Methicillin-resistant *Staphylococcus aureus* (MRSA) outbreaks in hospitals is a worldwide problem. In addition, conventional approaches that distinguish between MRSA and methicillin-sensitive *S. aureus* (MSSA) are slow, expensive, and time-consuming [40]. Hence, the rapid differentiation of MRSA from MSSA is essential for appropriate therapy and intervention for cross-infection control [41]. Edward-Jones et al. reported the spectra obtained from MRSA and MSSA strains and revealed that some peaks were specific to MRSA [42]. More recently, changes in the proteome of *Candida albicans*, corresponding to its fluconazole resistance have been reported [43].

Currently, MALDI-TOF MS identification warrants research to obtain bacterial colonies for the acquisition of spectra, as it cannot directly identify pathogens in a sample. Thus, novel enrichment techniques coupled with MALDI-TOF MS should be explored, as MALDI-TOF MS—based methods of antibiotic susceptibility require further development.

Recently, an exciting and novel approach, MALDI-TOF MS—based resistance test with stable isotopes (MS-RESIST), was described [44]. The use of MALDI-TOF MS profiling in combination with a growth medium containing isotopically labeled "heavy" amino acids facilitates the detection of resistant microorganisms in 3 hours or less directly from a profile spectrum. In the MS-RESIST method, growing microorganisms incorporate heavy amino acids into proteins, thereby increasing their mass and resulting in detectable shifts in protein m/z values [44]. In the presence of antibiotics, only resistant microorganisms can grow and incorporate the labeled heavier amino acids, leading to differences in the mass spectra of susceptible and resistant isolates, allowing their differentiation.

7.9 CONCLUSION

MALDI-TOF MS identification of bacteria at the species level remains unsatisfactory, with the primary problem being an incomplete database. Perhaps, an augmentation of the commercial database by incorporating mass spectra obtained in-house from local clinical isolates might increase the identification. A routine prospective study to assess whether the augmented database can enhance the performance of MALDI-TOF MS for routine identification of bacteria indicated that refinement of a commercial database can be achieved relatively quickly and effectively by incorporating MS spectra of clinical isolates obtained in a clinical laboratory. Bacterial identification using MALDI-TOF MS is now widely used in the real-world medicine, representing a revolutionary shift in clinical diagnostic microbiology. Admittedly, rapid identification of causative microbes is imperative for appropriate patient care in cases of bloodstream infections.

REFERENCES

[1] Nomura F. Proteome-based bacterial identification using matrix-assisted laser desorption ionization-time of flight mass spectrometry (MALDI-TOF MS): a revolutionary shift in clinical diagnostic microbiology. Biochim Biophys Acta 2015;1854:528—37.

[2] Clark AE, KaletaEJ, Arora A, Wolk DM. Matrix-assisted laser desorption ionization-time of flight mass spectrometry: a fundamental shift in the routine practice of clinical microbiology. Clin Microbiol Rev 2013;26:547−603.

[3] Carbonnelle E, Mesquita C, Bille E, Day N, Dauphin B, Beretti JL, et al. MALDI-TOF mass spectrometry tools for bacterial identification in clinical microbiology laboratory. Clin Biochem 2011;44:104−9.

[4] Croxatto A, Prod'hom G, Greub G. Applications of MALDI-TOF mass spectrometry in clinical diagnostic microbiology. FEMS Microbiol Rev 2012;36:380−407.

[5] Sogawa K, Watanabe M, Sato K, Segawa S, Ishii C, Miyabe A, et al. Use of the MALDI BioTyper system with MALDI-TOF mass spectrometry for rapid identification of microorganisms. Anal Bioanal Chem 2011;400:1905−11.

[6] Tan KE, Ellis BC, Lee R, Stamper PD, Zhang SX, Carroll KC. Prospective evaluation of a matrix-assisted laser desorption ionization-time of flight mass spectrometry system in a hospital clinical microbiology laboratory for identification of bacteria and yeasts: a bench-by-bench study for assessing the impact on time to identification and cost-effectiveness. J Clin Microbiol 2012;50:3301−8.

[7] Cherakaoui A, Emonet S, Fernandez J. Evaluation of matrix-assisted laser desorption ionization-time of flight mass spectrometry for rapid identification of beta-hemolytic streptococci. J Clin Microbiol 2011;49:3004−5.

[8] Segawa S, Sawai S, Murata S, Nishimura M, Beppu M, Sogawa K, et al. Direct application of MALDI-TOF mass spectrometry to cerebrospinal fluid for rapid pathogen identification in a patient with bacterial meningitis. Clin Chim Acta 2014;435:59−61.

[9] Segawa S, Nishimura M, Sogawa K, Tsuchida S, Murata S, Watanabe M, et al. Identification of Nocardia species using matrix-assisted laser desorption/ionization-time-of-flight mass spectrometry. Clin Proteomics 2015;12:6.

[10] Tanner H, Evans JT, Gossain S, Hussain A. Evaluation of three sample preparation methods for the direct identification of bacteria in positive blood cultures by MALDI-TOF. BMC Res Notes 2017;10(1):48.

[11] Christner M, Rohde H, Wolters M, Sobottka I, Wegscheider K, Aepfelbacher M. Rapid identification of bacteria from positive blood culture bottles by use of matrix-assisted laser desorption-ionization time of flight mass spectrometry fingerprinting. J Clin Microbiol 2010;48:1584−91.

[12] Sandalakis V, Goniotakis I, Vranakis I, Chochlakis D, Psaroulaki A. Use of MALDI-TOF mass spectrometry in the battle against bacterial infectious diseases: recent achievements and future perspectives. Expert Rev Proteomics 2017;22:1−15.

[13] Kaleta EJ, Clark AE, Cherakaoui A, Wysocki VH, Ingram EL, Schrenzel J, et al. Comparative analysis of PCR-electrospray ionization/mass spectrometry (MS) and MALDI-TOF/MS for the identification of bacteria and yeast from positive blood culture bottles. Clin Chem 2011;57:1057−67.

[14] Calderaro A, Arcangeletti MC, Rodighiero I, Buttrini M, Montecchini S, Vasile Simone R, et al. Identification of different respiratory viruses, after a cell culture step, by matrix assisted laser desorption/ionization time of flight mass spectrometry (MALDI-TOF MS). Sci Rep 2016;6:36082.

[15] Chen JH, Ho PL, Kwan GS, She KK, Siu GK, Cheng VC, et al. Direct bacterial identification in positive blood cultures by use of two commercial matrix-assisted laser desorption ionization−time of flight mass spectrometry systems. J Clin Microbiol 2013;51:1733−9.

[16] Bazzi AM, Rabaan AA, El Edaily Z, John S, Fawarah MM, Al-Tawfiq JA. Comparison among four proposed direct blood culture microbial identification methods using MALDI-TOF MS. J Infect Public Health 2016;10:308−15.

[17] Prod'hom G, Bizzini A, Durussel C, Bille J, Greub G. Matrix-assisted laser desorption ionization-time of flight mass spectrometry for direct bacterial identification from positive blood culture pellets. J Clin Microbiol 2010;48:1481—3.

[18] Bernhardt M, Pennell DR, Almer LS, Schell RF. Detection of bacteria in blood by centrifugation and filtration. J Clin Microbiol 1991;29(3):422—5.

[19] Seng P, Drancourt M, Gouriet F, La Scola B, Fournier PE, Rolain JM, et al. Ongoing revolution in bacteriology: routine identification of bacteria by matrix-assisted laser desorption ionization time-of-flight mass spectrometry. Clin Infect Dis 2009;15(49):543—51.

[20] Anhalt JP, Fenselau C. Identification of bacteria using mass spectrometry. Anal Chem 1975;47:219—25.

[21] Krishnamurthy T, Ross PL. Rapid identification of bacteria by direct matrix-assisted laser desorption/ionization mass spectrometric analysis of whole cells. Rapid Commun Mass Spectrom 1996;10(15):1992—6.

[22] Claydon MA, Davey SN, Edwards-Jones V, Gordon DB. The rapid identification of intact microorganisms using mass spectrometry. Nat Biotechnol 1996;14(11):1584—6.

[23] Eigner U, Holfelder M, Oberdorfer K, Betz-Wild U, Bertsch D, Fahr AM. Performance of a matrix-assisted laser desorption ionization-time-of-flight mass spectrometry system for the identification of bacterial isolates in the routine clinical laboratory. Clin Lab 2009;55(7-8):289—96.

[24] Cherkaoui A, Hibbs J, Emonet S, Tangomo M, Girard M, Francois P, et al. Comparison of two matrix-assisted laser desorption ionization-time of flight mass spectrometry methods with conventional phenotypic identification for routine identification of bacteria to the species level. J Clin Microbiol 2010;48(4):1169—75.

[25] Buchan BW, Riebe KM, Ledeboer NA. Comparison of the MALDI Biotyper system using Sepsityper specimen processing to routine microbiological methods for identification of bacteria from positive blood culture bottles. J Clin Microbiol 2012;50:346—52.

[26] Martiny D, Dediste A, Vandenberg O. Comparison of an in-house method and the commercial Sepsityper kit for bacterial identification directly from positive blood culture broths by matrix-assisted laser desorption-ionisation time-of-flight mass spectrometry. Eur J Clin Microbiol Infect Dis 2012;31:2269—81.

[27] Meex C, Neuville F, Descy J, Huynen P, Hayette MP, De Mol P, et al. Direct identification of bacteria from BacT/ALERT anaerobic positive blood cultures by MALDI-TOF MS: MALDI Sepsityper kit versus an in-house saponin method for bacterial extraction. J Med Microbiol 2012;61:1511—16.

[28] Bidart M, Bonnet I, Hennebique A, Kherraf ZE, Pelloux H, Berger F, et al. An in-house assay is superior to Sepsityper for direct matrix-assisted laser desorption ionization-time of flight (MALDI-TOF) mass spectrometry identification of yeast species in blood cultures. J Clin Microbiol 2015;53:1761—4.

[29] Ferreira L, Sanchez-Juanes F, Gonzalez-Avila M, Cembrero-Fucinos D, Herrero-Hernandez A, Gonzalez-Buitrago JM, et al. Direct identification of urinary tract pathogens from urine samples by matrix assisted laser desorption ionization-time of flight mass spectrometry. J Clin Microbiol 2010;48:2110—15.

[30] Hsieh SY, Tseng CL, Lee YS, Kuo AJ, Sun CF, Lin YH, et al. Highly efficient classification and identification of human pathogenic bacteria by MALDI-TOF MS. Mol Cell Proteomics 2008;7:448—56.

[31] Schubert S, Weinert K, Wagner C, Gunzl B, Wieser A, Maier T, et al. Novel, improved sample preparation for rapid, direct identification from positive blood cultures using matrix-assisted laser desorption/ionization time-of-flight (MALDI-TOF) mass spectrometry. J Mol Diagn 2011;13:701—6.

[32] Scott Jamie S, Sterling Sarah A, To Harrison, Seals Samantha R, Jones Alan E. Diagnostic performance of matrix-assisted laser desorption ionisation time-of-flight mass spectrometry in blood bacterial infections: a systematic review and meta-analysis. Infect Dis 2016;48(7):530—6.

[33] Íñigo M, Coello A, Fernández-Rivas G, Rivaya B, Hidalgo J, Quesada MD, et al. Direct identification of urinary tract pathogens from urine samples, combining urine screening methods and matrix-assisted laser desorption ionization-time of flight mass spectrometry. J Clin Microbiol 2016;54(4):988—93.

[34] Veron L, Mailler S, Girard V, Muller BH, L'Hostis G, Ducruix C, et al. Rapid urine preparation prior to identification of uropathogens by MALDI-TOF MS. Eur J Clin Microbiol Infect Dis 2015;34(9):1787—95.

[35] Rajagopal MU, Hathout Y, MacDonald TJ, Kieran MW, Gururangan S, Blaney SM, et al. Proteomic profiling of cerebrospinal fluid identifies prostaglandin D2 synthase as a putative biomarker for pediatric medulloblastoma: a pediatric brain tumor consortium study. Proteomics 2011;11(5):935—43.

[36] Samuel N, Remke M, Rutka JT, Raught B, Malkin D. Proteomic analyses of CSF aimed at biomarker development for pediatric brain tumors. J Neurooncol 2014;118:225—38.

[37] Núñez Galindo A, Kussmann M, Dayon L. Proteomics of cerebrospinal fluid: throughput and robustness using a scalable automated analysis pipeline for biomarker discovery. Anal Chem 2015;87(21):10755—61.

[38] Chiasserini D, van Weering JR, Piersma SR, Pham TV, Malekzadeh A, Teunissen CE, et al. Proteomic analysis of cerebrospinal fluid extracellular vesicles: a comprehensive dataset. J Proteomics 2014;106:191—204.

[39] Maurer MH, Berger C, Wolf M, Futterer CD, Feldmann Jr RE, Schwab S, et al. The proteome of human brain microdialysate. Proteome Sci 2003;1:7.

[40] Spencer RC. Predominant pathogens found in the European prevalence of infection in intensive care study. Eur J Clin Microbiol Infect Dis 1996;15:281—5.

[41] Archer GL. Staphylococcus aureus: a well-armed pathogen. Clin Infect Dis 1998;26:1179—81.

[42] Edwards-Jones V, Claydon MA, Evason DJ, Walker J, Fox AJ, Gordon DB. Rapid discrimination between methicillin-sensitive and methicillin-resistant Staphylococcus aureus by intact cell mass spectrometry. J Med Microbiol 2000;49:295—300.

[43] Marinach C, Alanio A, Palous M, Kwasek S, Fekkar A, Brossas JY, et al. MALDI-TOF MS-based drug susceptibility testing of pathogens: the example of *Candida albicans* and fluconazole. Proteomics 2009;9:4627—31.

[44] Sparbier K, Lange C, Jung J, Wieser A, Schubert S, Kostrzewa M. MALDI biotyper based rapid resistance detection by stable-isotope labeling. J Clin Microbiol 2013;51:3741—8.

CHAPTER EIGHT

Detecting Bacterial Resistance, Biomarkers, and Virulence Factors by MALDI-TOF Mass Spectrometry

María Dolores Rojo-Martín
Department of Microbiology, University Hospital Virgen de las Nieves, Granada, Spain

8.1 MATRIX-ASSISTED LASER DESORPTION IONIZATION—TIME OF FLIGHT MASS SPECTROMETRY FOR DETECTION OF ANTIBIOTIC RESISTANCE

One of the most important applications of matrix-assisted laser desorption ionization—time of flight (MALDI-TOF) mass spectrometry is in the diagnostic laboratory, as it enables the rapid, effortless, cost-effective, and accurate identification of bacteria and fungi. If this technology also allowed determining antimicrobial resistance, its contribution would be a revolution.

Infections by multidrug-resistant bacteria (MDRB) are a major problem in hospitals. More specifically, carbapenemase-producing Gram-negative bacilli are of special concern, due to their rapid spread and the few therapeutical options available. Early detection and antibiotic resistance profiling are crucial to controlling MDRB infection, as they allow the administration of the appropriate therapy and the establishment of specific infection control measures. Microbiological detection of MDRB is based on phenotypic testing, which takes about 18 hours. However, the sensitivity and specificity of phenotypic tests are often insufficient and fail to detect certain mechanisms of resistance. Genotype-based methods are more rapid, sensitive, and specific, but they require the use of specific primers for each of the enzymes analyzed, which makes it a labor-intensive and expensive method.

In this context, MALDI-TOF emerges as a plausible alternative to face the challenges of detecting MDRB in clinical laboratories.

The Use of Mass Spectrometry Technology (MALDI-TOF) in Clinical Microbiology.
DOI: https://doi.org/10.1016/B978-0-12-814451-0.00008-3

8.1.1 Detection of resistance mechanisms based on enzymatic degradation

8.1.1.1 Detection of β-lactamases

Progress has been significant in the identification of resistance mechanisms by MALDI-TOF mass spectrometry based on the enzymatic degradation of antimicrobial agents, more specifically, in the detection of β lactamases.

The β-lactam antibiotics after β-lactam ring hydrolysis by β lactamases generate different mass products resulting from the addition of a water molecule (an 18-Da mass increase). Also, some β-lactam antimicrobials are decarboxylated. MALDI-TOF mass spectrometry of the original molecule shows peaks in its spectrum (protein fingerprint) that correspond to β-lactam and its salts. In contrast, the spectra obtained after incubation of a bacterial suspension with antibiotic (1−3 hours)—in the case of hydrolysis—will cause the loss of β-lactam peaks, or only the peaks of its degradation products will be visible. This method can be used for detecting carbapenemases in *Enterobacteriaceae*, *Pseudomonas aeruginosa*, and *Acinetobacter baumannii*.

For the carbapenem hydrolysis assay, all studies follow the same methodology, namely, a fresh bacterial culture is suspended in a buffer and centrifuged (some protocols skip this step and the bacterial colony are directly suspended in a reaction buffer with antibiotic). The pellet is resuspended in a reaction buffer containing the β-lactam molecule. After incubation at 35°C for 1−3 hours, the reaction mixture is centrifuged and the supernatant is mixed with a proper matrix, and measured by MALDI-TOF. Spectra showing peaks of β-lactam, its salts (generally sodium salts), and/or its degradation products are analyzed using appropriate software usually provided by the manufacturer.

It should be noted that matrix is also visible in spectra. Thus, α-cyano-4-hydroxycinnamic acid (CHCA) produces a peak of 380 Da, which can exceed the peak of meropenem (384.5 Da), thereby hindering the visibility of the latter, its salts, and degradation products in spectrum. To avoid this problem, some authors use a different matrix—dihydroxybenzoic acid (DHB)—which does not show peaks in the 375−475 Da, which is the range of peaks of carbapenem and its degradation products. Another strategy is used in the hydrolysis assay, imipenem (peak: 300 Da) or ertapenem (peak: 476.5 Da, monosodium salt peak: 498.5 Da) instead of meropenem, which produce peaks that do not overlap with those of CHCA (first-choice matrix over DHB, which produces heterogeneous preparations complicating the acquisition of the spectra).

In one of the first studies performed to detect carbapenemases by MALDI-TOF, Hrabak et al. [1] detected meropenem and its degradation products using this technology. The authors used a 20 mM Tris-HCl (pH 6.8) buffer and an incubation time of 3 hours. Of the different matrices tested, DHB revealed as the best option. The assay was validated with 124 strains. Thirty isolates produced different carbapenemases (VIM, IMP, NDM-1, and KPC-2) and were identified as *P. aeruginosa* and different species of *Enterobacteriaceae*. They considered a positive result the disappearance in the spectrum of 383 Da (meropenem molecule) and 405 Da (meropenem sodium salt) peaks. A sensitivity of 96.67% and a specificity of 97.87% were obtained. The specificity of the assay was improved later by the same authors [2] by adding 0.01% sodium dodecyl sulfate (SDS) to the reaction buffer. SDS reduces the amount of bacterial cells, decreases incubation time to 2 hours, and improves the visualization of meropenem degradation products. This method was validated on 108 carbapenemase-producing species of *Enterobacteriaceae*, two NDM-1-producing *A. baumannii* isolates, and 35 carbapenem-resistant enterobacteria producing no carbapenemase.

Burckhardt and Zimmermann [3] validated a similar method based on ertapenem for the detection of carbapenemases in enterobacteria. First, the authors determined the characteristic mass spectrum of pure ertapenem and found that it showed four peaks, namely, 450 Da (hydrolyzed and decarboxylated ertapenem without sodium), 476 Da (ertapenem without sodium), 498 Da (monosodium salt), and 521 Da (disodium salt). CHCA was used as matrix. A result was interpreted as positive for carbapenemase production if the peaks for ertapenem (476, 498, and 521 Da) disappeared completely during the incubation time. Next, hydrolysis assays were conducted at different incubation times with KPC-2, NDM-1, IMP, and VIM-producing enterobacteria. The authors noted that the time of incubation required to degrade ertapenem differed according to the enzyme carried: NDM-1 and IMP-1 (1 hour), IMP-2, KPC-2, VIM-1 (1.5 hours), and VIM-2 (2.5 hours). A 100% sensitivity and specificity were achieved.

Kemp et al. detected the production of carbapenemases in *A. baumannii* by imipenem hydrolysis [4] using a panel of 106 strains of *A. baumannii*: 63 carbapenemase-producing and 43 non-carbapenemase-producing strains (7 carbapenem-resistant and 36 carbapenem-susceptible strains). A result was positive for carbapenemase production if the peak for imipenem at 300 Da disappeared during the incubation time and the peak of the natural metabolite at 254 Da increased so that the resulting imipenem to metabolite ratio was <0.5. This assay showed a sensitivity and specificity of 100.0% for an incubation time of 4 hours.

Álvarez-Buylla et al. [5] developed an optimized protocol for the detection of carbapenemases in *Acinetobacter* species by MALDI-TOF mass spectrometry. Initially, the authors established the characteristic spectrum of imipenem and selected peaks at 300 and 489 mass to charge ratio (m/z) to identify the presence of carbapenemases. Assays were performed at different concentrations of bacterial inoculum and imipenem in distinct buffers at a variety of incubation times. The best option was a bacterial inoculum of 2.5×10^{10} CFU/mL, 1 mg/mL of imipenem, ClNa 0.45% Tris-HCl 20 mM buffer with 1 hour of incubation at 35°C, under constant shaking (500 rpm). After 1 hour of incubation, all carbapenemase-producing strains caused a significant decrease in the intensity of peaks at 300 and 489 Da in the spectrum. By the addition of dipicolinic acid, the authors could distinguish between metallo-β-lactamase (MBL)-producing strains (inhibition of imipenem hydrolysis) and oxacillinase-producing strains (imipenem hydrolysis was not inhibited).

Sparbier et al. [6] used MALDI-TOF mass spectrometry to detect a variety of β-lactam antimicrobials (ampicillin, piperacillin, cefotaxime, ceftazidime, ertapenem, imipenem, and meropenem) and their degradation products after hydrolysis, with a sensitivity and specificity of 100%. The authors used *Escherichia coli* DH5α (ATCC) as negative control, five β-lactamase-producing *E. coli* strains, two carbapenemase-producing *Klebsiella pneumoniae* strains, and 1 carbapenemase-negative *K. pneumoniae* strain. They identified the peaks corresponding to each of the molecules and defined the resistance and sensitivity patterns for each one of them. The authors concluded that for a reliable analysis, it is not only the loss of peaks of the original molecule but also the appearance of peaks of hydrolysis what has to be monitored. Strains were classified as presumably susceptible if the intensity distributions of the nonhydrolyzed and the hydrolyzed forms were similar to those for the negative control. Strains were classified as resistant if the intensities of the hydrolyzed forms represented 80% or more of the intensities of the nonhydrolyzed and the hydrolyzed forms. To identify the β-lactamase type, assays were performed using clavulanic acid (penicillinases, ESBL), tazobactam, and aminophenylboronic acid (KPC).

Hoyos et al. [7] selectively inhibited ertapenem hydrolysis with EDTA to identify MBLs. The study included 49 non-carbapenemase-producing carbapenem-resistant and 14 carbapenemase-producing clinical strains: six

Enterobacteriaceae (two IMP, three VIM, and one KPC), and eight *Pseudomonas aeruginosa* (seven VIM and one IMP). Ertapenem hydrolysis by carbapenemases showed no peak corresponding to the non-hydrolyzed forms: 476.5 Da [M + H] +, 498.5 Da [M + Na] +, and 520.5 Da [M + 2Na] + and the appearance of degradation products at 450.5 Da $[M_{hydr./decarb.} + H]^+$, 494.5 Da $[M_{hydr.} + H]^+$, and 472.5 Da $[M_{hydr./decarb.} + Na]^+$ (Fig. 8.1). When EDTA was added to the same assay, ertapenem hydrolysis was inhibited, which enabled the identification of MBL-producing strains.

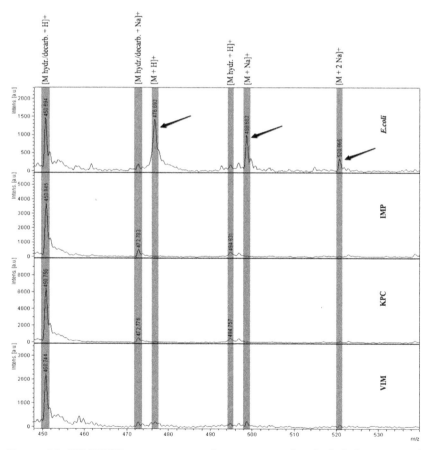

Figure 8.1 MALDI-TOF mass spectra of ertapenem after hydrolysis assay with *Escherichia coli* ATCC 25922 and carbapenemase-producing strains.

8.1.1.2 Detection of β-lactamases directly from blood culture

The use of MALDI-TOF for bacterial identification directly from positive blood cultures has an important impact in the treatment of patients with Gram-negative bacteremia [8]. However, the increased incidence of multidrug-resistant Gram-negative bacteria makes it difficult to predict resistance patterns. Another application of MALDI-TOF is for the direct detection of β-lactamases in blood culture. This is of paramount importance for the appropriate treatment of patients with Gram-negative bacteremia. Excellent results have been reported in a range of studies [6,9,10]. Thus, Sparbier et al. [6] performed an assay with 1 mL of blood from a blood culture inoculated with KPC-producing *K. pneumoniae* and were able to hydrolyze a suspension of ertapenem after 3 hours of incubation using the MALDI Sepsityper kit (Bruker Daltonik GmbH, Germany). Jung et al. [9] carried out a similar assay in blood cultures of patients with Gram-negative bacteremia (100 consecutive blood cultures containing *Enterobacteriaceae* for resistance against third-generation cephalosporins). The results of the β-lactamase assay were compared with those of conventional methods. The assay permitted discrimination between *E. coli* strains that were resistant or susceptible to aminopenicillins, with sensitivity and specificity of 100%. The same was true for resistance to third-generation cephalosporins in *Enterobacteriaceae* that constitutively produced Class C-lactamases. Discrimination was more difficult in species expressing Class A lactamases, with a sensitivity and specificity of 100% and 91.5%, respectively. The test permitted the prediction of resistance within 2.5 hours after blood culture was found positive. Carvalhaes et al. [10] detected KPC carbapenemases (*K. pneumoniae and Enterobacter cloacae*) and SPM-1 *(P. aeruginosa)* in 4 hours using MALDI-TOF mass spectrometry. Sakarikou et al. [11] detected carbapenemase-producing *K. pneumoniae* strains (KPC, OXA, and VIM) from blood cultures after 30 minutes to 3 hours of incubation with an ertapenem solution, with a 100% level of agreement with conventional methods.

The evidence provided by these studies demonstrates that MALDI-TOF mass spectrometry is a rapid, reliable method for the detection and identification of a range of β-lactamases in few hours—especially carbapenemases—even directly from blood culture vials. It may contribute to faster readjustment of empirical antimicrobial therapy and implementation of infection control measures.

However, this methodology is flawed with some limitations, namely it does not detect other mechanisms of resistance against carbapenems such as alterations of porin and efflux pump overexpression in *K. pneumoniae* and *P. aeruginosa*, or PBP alterations in *A. baumannii*. Additionally, low expression of some carbapenemases can make it difficult to detect them. The protocols used in most clinical laboratories are pending validation. Therefore, results must be considered in combination with those of phenotypic and molecular tests. The interpretation and evaluation of the spectra require expert knowledge, as it is based on a qualitative visual examination. Also, new technologies for the automated interpretation of results are under development. An example is the MBT STAR-BL (MALDI Biotyper, Bruker) module for the automated detection of β-lactamases after a hydrolysis assay. This software automatically calculates the summed hydrolyzed and nonhydrolyzed antibiotic signal intensities and the corresponding ratio. Results are yielded in the form of a box plot diagram.

The principle of enzymatic degradation can be applied to other antimicrobials such as aminoglycosides. However, this methodology has not yet been validated.

8.1.2 Detection of resistance through bacterial protein profiling

This application of MALDI-TOF mass spectrometry is based on differences in protein fingerprint or spectrum (the set of peaks of proteins expressed by resistance genes) between resistant microorganisms and susceptible microorganisms of the same species. This technique has been successfully used in a number of studies performed in clinical laboratories, albeit it needs optimization and validation.

8.1.2.1 Detection of methicillin-resistant Staphylococcus aureus

Inconsistent results have been obtained in studies examining the ability of MALDI-TOF to detect methicillin-resistant *Staphylococcus aureus* (MRSA). One of the first studies in this field was undertaken by the University of Manchester in 2000 [12]. Assays were performed in seven nosocomial MRSA strains, and seven methicillin-susceptible *S. aureus* (MSSA) strains. The strains were analyzed by MALDI-TOF using intact bacterial cells and 5-chloro-2-mercapto-benzothiazole as the matrix. The authors detected that MRSA and MSSA produced 14 and 2 specific peaks, respectively, and observed that MRSA produced more peaks (82−209) in the spectrum than MSSA (37−67), which allowed visual discrimination. Similar results were obtained by Du et al. [13]. Bernardo et al. [14] studied well-characterized *S. aureus* strains (ATCC 29213 (MSSA), ATCC 43330 (MRSA), and nine clinical isolates). The authors performed 3-month monitoring of spectra after successive subcultures, with good reproducibility. A unique spectrum was not found for MRSA but for an individual strain which was conserved over time. Therefore, it was concluded that this technique can be useful for clonal characterization of *S. aureus* strains. Lu et al. [15] detected specific markers that made it possible to discriminate nosocomial MRSA (SCC*mec, Staphylococcal cassette chromosome mec,* I-III types) and community-acquired MRSA (SCC*mec,* IV–V types), but no unique spectra were identified for MRSA and MSSA. Szabados et al. [16] compared the spectra of two strains of the same genetic origin—an *SCCmec* carrier (methicillin-resistant) and an *SCCmec* non-carrier (methicillin-susceptible)—by MALDI-TOF and obtained virtually identical spectra. In an assay with 100 clinical strains, Wang et al. [17] distinguished two groups based on two greater intensity peaks (3.784 and 5.700 Da) identified for MRSA but not for MSSA.

In the light of the results obtained by these studies, the prediction of MRSA from MALDI-TOF spectra is challenging. Indeed, differences in spectra between MRSA and MSSA are probably due to variations in peaks related to different clonal lineages. Therefore, mass spectrometry does not seem to be reliable for the detection of methicillin resistance in *S. aureus*.

8.1.2.2 Detection of other mechanisms of resistance

The specific spectrum of some antimicrobial-resistant microorganisms has been successfully used for this purpose. Camara et al. [18] detected a 29-kDa-specific β-lactamase peak for ampicillin-resistant *E. coli* strains. Conversely, Schaumann et al. [19] were not able to find any difference between β-lactamase producers and nonproducers *Enterobacteriaceae* and *P. aeruginosa* isolates. In a recent study, a unique peak of 39,850 kDa was identified as specific of CMY-2-type cephalosporinase-producing Enterobacteria [20].

Cai et al. [21] identified two peaks of 38,000 and 19,000 Da as specific of the OmpK36 porin and detected the loss of this porin in eight carbapenem-resistant *K. pneumoniae* isolates and a *Klebsiella oxytoca* isolate.

Griffin et al. [22] successfully detected vanB-positive *Enterococcus faecium*. The *vanB* gene provides resistance to glycopeptides. Detection was performed using a statistical model based on the spectra obtained after ethanol–formic acid extraction. Prospective validation of results yielded a sensitivity and specificity of 96.7% and 98.1%, respectively. The efficacy of MALDI-TOF in discriminating vanA-positive *E. faecium* from vanB-negative *E. faecium* has also been assessed, with a sensitivity and specificity >90% [23]. Yet, as isolates were collected from the same geographical region, they could be clonally related, which compromises the results obtained.

Based on MALDI-TOF spectra, some authors could discriminate *cfiA*-positive *Bacteroides fragilis* from *cfiA*-negative *B. fragilis* (this gene encodes a metallo-β lactamase) [24,25]. Johansson et al. [26] detected *cfiA*-positive *B. fragilis* from blood cultures by MALDI-TOF.

8.1.3 Determination of minimal profile changing concentration

This application of MALDI-TOF mass spectrometry for the detection of antimicrobial resistance is more similar to phenotypic testing. This technique is based on the detection of the minimal antimicrobial concentration that causes a change in the protein fingerprint of the microorganism studied. This application has been mainly investigated for the assessment of yeast susceptibility to antifungal agents. Marinach et al. [27] compared the spectra of different yeast strains after 24-hour incubation in a liquid solution with different antifungal concentrations. Mass spectrometry protein fingerprint analysis revealed that a dramatic change occurred in protein spectrum at a specific concentration. The minimal fluconazole concentration causing a change in protein spectrum was called "minimal profile changing concentration" and showed a high correlation with minimum inhibitory concentration as determined by the standard Clinical and Laboratory Standards Institute method. The main drawback of this method is that yeast requires 18-hour incubation in antifungal solution. This technique has been proven to be effective [28] in assessing susceptibility to caspofungin in *Candida* spp. and *Aspergillus* spp. A study was recently published where incubation time was reduced to 3 hours and only three antifungal concentrations were used: without antifungal, at breakpoint, and at maximal concentration, with good results [29]. This is a rapid technique, with automated interpretation, that could be incorporated into clinical laboratory practice. However, the reproducibility and robustness of this method have yet to be proven.

8.1.4 MALDI-TOF-based resistance detection by stable isotope labeling

Recently, a new method based on MALDI-TOF has been developed for the detection of resistance by the use of culture media labeled with stable isotopes. Only resistant microorganisms can grow in the presence of antimicrobials. Resistant microorganisms incorporate isotopically labeled amino acids, increasing protein masses and thereby leading to mass shifts of their corresponding peaks in the profile spectra. This causes a change in the spectrum with respect to susceptible microorganisms. Demirev et al. [30] used as control a commercially available culture medium in which 98% of carbon atoms were ^{13}C. Streptomycin is added to determine the sensitivity of a range of *Bacillus* spp. and *E. coli* strains.

Sparbier et al. [31] demonstrated the efficacy of this method in detecting MRSA. A variety of software tools have been tested for the automated interpretation of results. The authors used 10 MRSA and 10 MSSA strains for the study, and 14 MRSA and 14 MSSA strains for validation. Each strain was incubated in 100 μL of three different media for 3 hours, as follows: Modified Dulbecco's Eagle "normal" (lysine 0.1 g/L), "heavy" ($^{13}C_6{}^{15}N_2$-L-lysine 0.1 g/L) and "heavy + OXA" ($^{13}C_6{}^{15}N_2$-L-lysine 0.1 g/L + oxacilline 60 mg/L). MRSA strains can grow in the presence of antimicrobials. Spectra showed peaks corresponding to the addition of isotope-labeled lysine, which does not occur in MSSA strains. This application was investigated in another study [32] to determine the susceptibility of *P. aeruginosa* against meropenem, ciprofloxacin, and tobramycin, with good results.

8.2 IDENTIFICATION OF DIAGNOSTIC SERUM BIOMARKERS FOR INFECTIOUS DISEASES BY MASS SPECTROMETRIC PROFILING

MALDI-TOF mass spectrometry has also been used to identify biomarkers of infectious diseases. Yet, this use is a technical challenge. Serum is composed of proteins such as albumin, immunoglobulins, or transferrin (almost 90% of total serum/plasma proteins) that hinder the detection and identification of proteins present at low levels, which are potential biomarkers. To detect these proteins, the proteins present at high levels have to be eliminated by effective fractionation and separation methods. The identification of biomarkers is based on the analysis of the spectrometric profile of serum from infected patients as compared to that of controls. This will enable to establish a biomarker pattern that allows discriminating between infected and noninfected patients [33]. This method has been used to identify diagnostic serum biomarkers of infections caused by parasites such as *Trypanosoma cruzi* [34], *Plasmodium falciparum* [35], or *Opisthorchis viverrini infection* [36]. Kardush et al. [37] compared the spectrometric profile of serum from mice infected and noninfected by *Schistosoma mansoni* to identify specific biomarkers of disease stage. A wide range of potential biomarkers were identified, as the spectrometric profile of serum from infected mice was substantially different from that of noninfected mice, and initial, acute, and chronic disease could be easily identified.

Nevertheless, this technology has not yet been optimized and protocols are pending validation. Therefore, MALDI-TOF can only be used to identify diagnostic serum biomarkers for infectious diseases in research laboratories.

8.3 DETECTION OF TOXINS AND MICROBIAL ANTIGENS

Analysis of spectra allows the identification of other types of relevant proteins in clinical and epidemiological terms [33]. Gagnaire et al. [38] used MALDI-TOF on intact cells to detect the delta-toxin of *S. aureus* (peak 3.005 ± 5 Da). This is an indicator of accessory gene regulator (agr) system function. The agr system regulates the expression of virulence factors such as some exotoxins and proteins involved in the adherence and formation of biofilm.

Mass spectrometry has been used to detect bacterial antigens directly from the sample using antigen-specific antibodies bound to magnetic particles. Following the antigen—antibody reaction, antigen protein detection is carried out by MALDI-TOF after antigen—antibody complex dissociation. This method has enabled the detection of *Clostridium difficile* toxin [39], enterotoxigenic *E. coli* thermostable and thermolabile toxins [40], or the antigen secreted to the broth culture medium by *Mycobacterium tuberculosis complex* [41] during growth. In the latter case, nanodiamonds generated by detonation (ultradispersed diamonds) were used to capture the antigen.

The identification of new antigens by MALDI-TOF mass spectrometry has also been used for the development of vaccines and rapid diagnostic tests [42—44].

REFERENCES

[1] Hrabák J, Walkova R, Študentova V, Chudačkova E, Bergerova T. Carbapenemase activity detection by matrix-assisted laser desorption ionization-time of flight mass spectrometry. J Clin Microbiol 2011;49:3222−7.

[2] Hrabak J, Študentova V, Walkova R, Zemlickova H, Jakubu V, Chudackova E, et al. Detection of NDM-1, VIM-1, KPC, OXA-48, and OXA-162 carbapenemases by matrix-assisted laser desorption ionization-time of flight mass spectrometry. J Clin Microbiol 2012;50:2441−3.

[3] Burckhardt I, Zimmermann S. Using matrix-assisted laser desorption ionization-time of flight mass spectrometry to detect carbapenem resistance within 1 to 2.5 hours. J Clin Microbiol 2011;49:3321−4.

[4] Kempf M, Bakour S, Flaudrops C, Berrazeg M, Brunel JM, Drissi M, et al. Rapid detection of carbapenem resistance in *Acinetobacter baumannii* using matrix-assisted laser desorption ionization-time of flight mass spectrometry. PLoS One 2012;7:1−7.

[5] Álvarez-Buylla A, Picazo JJ, Culebras E. Optimized method for Acinetobacter species carbapenemase detection and identification by matrix-assisted laser desorption ionization-time of flight mass spectrometry. J Clin Microbiol 2013;51:1589−92.

[6] Sparbier K, Schubert S, Weller U, Boogen C, Kostrzewa M. Matrix-assisted laser desorption ionization-time of flight mass spectrometry-based functional assay for rapid detection of resistance against beta-lactam antibiotics. J Clin Microbiol 2012;50:927−37.

[7] Hoyos-Mallecot Y, Cabrera J, Miranda C, Rojo MD, Liébana C, Navarro JM. MALDI-TOF MS, a useful instrument for differentiating metallo-β-lactamases in Enterobacteriaceae and Pseudomonas spp. Lett Appl Microbiol 2014;58:325−9.

[8] Clerc O, Prodhom G, Vogne C, et al. Impact of matrix-assisted laser desorption ionization time-of-flight mass spectrometry on the clinical management of patients with Gram-negative bacteremia: a prospective observational study. Clin Infect Dis 2013;56:1101−7.

[9] Jung JS, Popp C, Sparbier K, Lange C, Kostrzewa M, Schubert S. Evaluation of matrix-assisted laser desorption ionization−time of flight mass spectrometry for rapid detection of β-lactam resistance in *Enterobacteriaceae* derived from blood cultures. J Clin Microbiol 2014;52:924−30.

[10] Carvalhaes CG, Cayô R, Visconde MF, Barone T, Frigatto EAM, Okamoto D, et al. Detection of carbapenemase activity directly from blood culture vials using MALDI-TOF MS: a quick answer for the right decision. J Antimicrob Chemother 2014;69:2132−6.

[11] Sakarikou C, Ciotti M, Dolfa C, Angeletti S, Favalli C. Rapid detection of carbapenemase-producing *Klebsiella pneumoniae* strains derived from blood cultures by matrix-assisted laser desorption ionization−time of flight mass spectrometry (MALDI-TOF MS). BMC Microbiol 2017;17:54.

[12] Edwards-Jones V, Claydon MA, Evason DJ, Walker J, Fox AJ, Gordon DB. Rapid discrimination between methicillin-sensitive and methicillin-resistant *Staphylococcus aureus* by intact cell mass spectrometry. J Med Microbiol 2000;49:295−300.

[13] Du Z, Yang R, Guo Z, Song Y, Wang J. Identification of *Staphylococcus aureus* and determination of its methicillin resistance by matrix-assisted laser desorption/ ionization time-of-flight mass spectrometry. Anal Chem 2002;74:5487−91.

[14] Bernardo K, Pakulat N, Macht M, Krut O, Seifert H, Fleer S, et al. Identification and discrimination of *Staphylococcus aureus* strains using matrix-assisted laser desorption/ionization-time of flight mass spectrometry. Proteomics 2002;2:747−53.

[15] Lu JJ, Tsai FJ, Ho CM, Liu YC, Chen CJ. Peptide biomarker discovery for identification of methicillin-resistant and vancomycin-intermediate *Staphylococcus aureus* strains by MALDI-TOF. Anal Chem 2012;84:5685−92.

[16] Szabados F, Kaase M, Anders A, Gatermann SG. Identical MALDI TOF MS-derived peak profiles in a pair of isogenic *SCCmec*-harboring and *SCCmec*-lacking strains of *Staphylococcus aureus*. J Infect 2012;65:400−5.

[17] Wang YR, Chen Q, Cui SH, Li FQ. Characterization of *Staphylococcus aureus* isolated from clinical specimens by matrix assisted laser desorption/ionization time-of-flight mass spectrometry. Biomed Environ Sci 2013;26:430−6.

[18] Camara JE, Hays FA. Discrimination between wild-type and ampicillin-resistant *Escherichia coli* by matrix-assisted laser desorption/ionization time-of-flight mass spectrometry. Anal Bioanal Chem 2007;389:1633−8.

[19] Schaumann R, Knoop N, Genzel GH, Losensky K, Rosenkranz C, Stîngu CS, et al. A step towards the discrimination of beta-lactamase-producing clinical isolates of Enterobacteriaceae and *Pseudomonas aeruginosa* by MALDI-TOF mass spectrometry. Med Sci Monit 2012;18:MT71−7.

[20] Papagiannitsis CC, Kotsakis SD, Tuma Z, Gniadkowski M, Miriagou V, Hrabak J. Identification of CMY-2-type cephalosporinases in clinical isolates of Enterobacteriaceae by MALDI-TOF MS. Antimicrob Agents Chemother 2014;58:2952−7.

[21] Cai JC, Hu YY, Zhang R, Zhou HW, Chen GX. Detection of OmpK36 porin loss in *Klebsiella spp.* by matrix-assisted laser desorption ionization-time of flight mass spectrometry. J Clin Microbiol 2012;50:2179−82.

[22] Griffin PM, Price GR, Schooneveldt JM, Schlebusch S, Tilse MH, Tess Urbanski T, et al. Use of matrix-assisted laser desorption ionization-time of flight mass spectrometry to identify vancomycin-resistant enterococci and investigate the epidemiology of an outbreak. J Clin Microbiol 2012;50:2918−31.

[23] Nakano S, Matsumura Y, Kato K, Yunoki T, Hotta G, Noguchi T, et al. Differentiation of vanA-positive *Enterococcus faecium* from vanA-negative *E. faecium* by matrix-assisted laser desorption/ionisation time-of-flight mass spectrometry. Int J Antimicrob Agents 2014;44:256−9.

[24] Wybo A, De Bel O, Soetens F, Echahidi K, Vandoorslaer M, Cauwenbergh Van, et al. Differentiation of *cfiA*-negative and *cfiA*-positive *Bacteroides fragilis* isolates by matrix-assisted laser desorption ionization-time of flight mass spectrometry. J Clin Microbiol 2011;49:1961−4.

[25] Nagy E, Becker S, Sóki J, Urbán E, Kostrzewa M. Differentiation of division I (*cfiA*-negative) and division II (*cfiA*-positive) *Bacteroides fragilis* strains by matrix-assisted laser desorption/ionization time-of-flight mass spectrometry. J Med Microbiol 2011;60:1584−90.

[26] Johansson A, Nagy E, Sóki J. Instant screening and verification of carbapenemase activity in Bacteroides fragilis in positive blood culture, using matrix-assisted laser desorption ionization-time of flight mass spectrometry. J Med Microbiol 2014;63:1105−10.

[27] Marinach C, Alanio A, Palous M, Kwasek S, Fekkar A, Brossas JY, et al. MALDI-TOF MS-based drug susceptibility testing of pathogens: the example of *Candida albicans* and fluconazole. Proteomics 2009;9:4627−31.

[28] De Carolis E, Vella A, Florio AR, Posteraro P, Perlin DS, Sanguinetti M, et al. Use of matrix-assisted laser desorption ionization-time of flight mass spectrometry for caspofungin susceptibility testing of *Candida* and *Aspergillus* species. J Clin Microbiol 2012;50:2479−83.

[29] Vella A, De Carolis E, Vaccaro L, Posteraro P, Perlin DS, Kostrzewa M, et al. Rapid antifungal susceptibility testing by matrix-assisted laser desorption ionization-time of flight mass spectrometry analysis. J Clin Microbiol 2013;51:2964−9.

[30] Demirev PA, Hagan NS, Antoine MD, Lin JS, Feldman AB. Establishing drug resistance in microorganisms by mass spectrometry. J Am Soc Mass Spectrom 2013;24:1194—201.

[31] Sparbier K, Lange C, Jung J, Wieser A, Schubert S, Kostrzewa M. MALDI Biotyper based rapid resistance detection by stable isotope labelling. J Clin Microbiol 2013;51:3741—8.

[32] Jung JS, Eberl T, Sparbier K, Lange C, Kostrzewa M, Schubert S, et al. Rapid detection of antibiotic resistance based on mass spectrometry and stable isotopes. Eur J Clin Microbiol Infect Dis 2014;33:949—55.

[33] Vila J, Zboromyrska Y, Burillo A, Bouza E. Perspectivas de futuro de la espectrometría de masas en microbiología. Enferm Infecc Microbiol Clin 2016;34 (Suppl 2):53—8.

[34] Ndao M, Spithill TW, Caffrey R, Li H, Podust VN, Perichon R, et al. Identification of novel diagnostic serum biomarkers for Chagas' disease in asymptomatic subjects by mass spectrometric profiling. J Clin Microbiol 2010;48:1139—49.

[35] Thezenas ML, Huang H, Njie M, Ramaprasad A, Nwakanma DC, Fischer R, et al. PfHPRT: a new biomarker candidate of acute *Plasmodium falciparum* infection. J Proteome Res 2013;12:1211—22.

[36] Khoontawad J, Laothong U, Roytrakul S, Pinlaor P, Mulvenna J, Wongkham C, et al. Proteomic identification of plasma protein tyrosine phosphatase alpha and fibronectin associated with liver fluke, *Opisthorchis viverrini*, infection. PLoS One 2012;7:e45460.

[37] Kardoush MI, Ward BJ, Ndao M. Identification of candidate serum biomarkers for Schistosoma mansoni infected mice using multiple proteomic platforms. PLoS One 2016;11:e0154465. Available from: https://doi.org/10.1371/journal.pone.0154465.

[38] Gagnaire J, Dauwalder O, Boisset S, Khau D, Freydiere AM, Ader F, et al. Detection of *Staphylococcus aureus* delta-toxin production by whole-cell MALDI-TOF mass spectrometry. PLoS One 2012;7:e40660.

[39] Kiyosuke M, Kibe Y, Oho M, Kusaba K, Shimono N, Hotta T, et al. Comparison of two types of matrix-assisted laser desorption/ionization time-of-flight mass spectrometer for the identification and typing of *Clostridium difficile*. J Med Microbiol 2015;64:1144—50.

[40] Kuo FY, Chang BY, Wu CY, Mong KK, Chen YC. Magnetic nanoparticle-based platform for characterization of Shiga-like toxin 1 from complex samples. Anal Chem 2015;87:10513—20.

[41] Soo PC, Kung CJ, Horng YT, Chang KC, Lee JJ, Peng WP. Detonation nanodiamonds for rapid detection of clinical isolates of *Mycobacterium tuberculosis* complex in broth culture media. Anal Chem 2012;84:7972—8.

[42] Boamah D, Kikuchi M, Huy NT, Okamoto K, Chen H, Ayi I, et al. Immunoproteomics identification of major IgE and IgG4 reactive *Schistosoma japonicum* adult worm antigens using chronically infected human plasma. Trop Med Health 2012;40:89—102.

[43] Dea-Ayuela MA, Rama-Iñiguez S, Bolás-Fernández F. Proteomic analysis of antigens from *Leishmania infantum* promastigotes. Proteomics 2006;6:4187—94.

[44] Wongkamchai S, Chiangjong W, Sinchaikul S, Chen ST, Choochote W, Thongboonkerd V. Identification of *Brugia malayi* immunogens by an immunoproteomics approach. J Proteomics 2011;74:1607—13.

CHAPTER NINE

Direct Identification of Pathogens From Blood Cultures by MALDI-TOF Technology

Elena Cuadros
Microbiology Hospital Universitario Virgen de las Nieves, Granada, Spain

9.1 INTRODUCTION

Nowadays, bloodstream infections (BSIs) still represent a major cause of morbidity and mortality for hospitalized patients worldwide [1−6]. Even though diagnosis and treatment have greatly improved over the last 40 years, it is critical to know that each hour of delay in initiating an appropriate antimicrobial therapy has been associated with a 7.6% decrease in survival for a septic patient who remains untreated or receives inappropriate antimicrobial therapy within the first 24 hours [7]; this fact has shown to be the single most powerful risk factor associated with mortality [8−10]. BSIs suppose as well an important load to the healthcare system, mostly due to increased length of stay that implies the largest use of antibiotics, a major risk of nosocomial infection and several additional hospital charges [11].

The management of BSI is based on two important parameters: the detection and identification of viable pathogen(s), and the antimicrobial susceptibility testing (AST) results. Therefore, a prompt and accurate identification of microorganisms from patients with bacteremia is the underlying function of any clinical microbiology laboratory due to the critical implication on the administration of the appropriate antimicrobial therapy, lowering costs, and improving patient outcomes [12−14]. Regarding this fact is also important to consider the implication of establishing empirical antimicrobial therapy. Up to 65% of intensive care patients showing symptoms of sepsis are already under presumptive antimicrobial treatment [2,15], which is chosen on the basis of clinical and

127

epidemiological data but usually involves the choice of broad-spectrum antibiotics that may favor the selection and spread of resistant pathogens, and increase the frequency of invasive fungal infections [3,15,16].

Plasma represents up to 55% of the total blood volume and offers a good opportunity to recover grown bacteria, as they are mostly present in the extracellular compartment [2,17]. For this purpose, blood culture (BC) is the gold standard on the diagnosis of BSI, used to discriminate whether a febrile episode is due to an infectious or a noninfectious cause and to provide information about the pathogen [18].

Since the beginning of automated BC techniques, conventional identification is based on time-consuming procedures that burden the proper management of the antibiotic therapy and supportive treatments.

Routine confirmation of microbial identification has depended on assays established into stages:

1. Early diagnostic and therapeutic decisions based on stain methods for microscopic morphology classification. The notification of Gram stain results in less than 1 hour influences the timely selection of an appropriate antimicrobial therapy, associated with a significant reduction of mortality in patients with BSI, compared with delayed process [4,5,19−22].
2. Confirmation of the pathogen by culture and growth characteristics. Performance of biochemical or antigenic techniques for the subsequent metabolic and phenotypic analysis of the microorganism.
3. AST to confirm therapeutic options and manage the treatment [6,8,13].

Keeping in mind the culture time for bacteria to grow to a detectable level, by an automated instrument, is approximately of 12−24 hours of incubation for Gram-negative, 24−48 hours for Gram-positive microorganisms, and that longer incubation times and culture medium additives may be required for fastidious or atypical microorganisms. It is easy to understand the implication of this delay on the report of the final complete identification of the pathogen [2,3,23]. Improvements in culture media and detection of growth procedures reduced partially the implied delay in results, the most recent generation of automates can detect even small bacterial growth [24−32].

During the last three decades of the previous century had been introduced serological, molecular and mass spectrometry techniques, replacing some of the several culture-based technologies used before [33−35]. When growth is detected by the automate, it is possible to perform direct identification of the pathogen by universal amplification and sequencing [36,37], nucleic acid—based fluorescence hybridization probes, such as fluorescence in situ hybridization [38—41], DNA microarrays [42], or molecular detection amplification and specific probes [43—47]. These procedures are performed directly from the positive BC and do not need subculture, but the number of pathogens that can be detected is limited. Moreover, being highly efficient the expensive cost and the requirement of high-qualified bacteriology technician make it difficult for these techniques to be set up by every clinical microbiology laboratory.

Likewise along the last decade matrix-assisted laser desorption ionization—time of flight (MALDI-TOF) mass spectrometry (MS) is being used in several clinical microbiology laboratories as routine technique applied directly to positive BC bottles, providing a rapid and reliable identification of the causal pathogen within hours; compared with the biochemical methods of identification for bacteria and yeasts from positive BCs, the use of a MALDI-TOF-based technique saves between 6 and 12 hours for Enterobacteriaceae and up to 48 hours for anaerobic bacteria [48].

9.2 MS BASICS APPLIED TO BCS

Mainly ribosomal proteins along with few housekeeping proteins, both with a mass range of 2—20 kDa, are the target of this technique because ribosomal proteins represent about 60%—70% of the dry weight of a microbial cell and demonstrate high species diversity, which perfectly serves the requirements for species identification and typing [49—52].

Previous to the MS the sample obtained is mixed directly on a stainless steel target plate with a saturated solution of a low-mass organic compound (matrix). Upon drying at room temperature, the matrix co-crystallizes with the sample. Once the target plate is inside the instrument, the sample within the matrix is sublimated and ionized by a pulsed UV laser beam (usually, an N_2 laser beam with a wavelength of 337 nm in commercial instruments), and the system includes, as well, a beam attenuator to adjust the irradiance [5] yielding the soft ionization.

The laser beam is focused on a small spot of the matrix-clinical sample crystalline surface (typically 0.05—0.2 mm in diameter), where peptides are then transformed into ions by either addition or loss of protons. The matrix is essential at this point as it acts both as a stage by which ionization can take place and as a proton supplier [5]. Soft ionization leads a short irradiation time that does not cause a significant loss of sample integrity [53−56]. Those ionized molecules are then accelerated at a fixed potential, which separates them according to their mass to charge ratio (m/z). Finally, the m/z ratios are detected by a TOF analyzer that defines the time required for each ionized molecule to travel the length of the flight tube (Fig. 9.1).

Based on the TOF information, a characteristic spectrum called peptide mass fingerprint (PMF) is generated. This spectrum reflects the mass peak profile specific for each bacterial or yeast species [57]. Identification is done by either comparing the PMF of an unknown organism with the PMFs contained in the database or by matching the masses of biomarkers of the unknown organism with the proteome database of bacterial and fungal peak profiles within seconds [58]. Databases are regularly updated increasing periodically the number of species incorporated, including molds and mycobacteria.

When the profiles obtained do not give a reliable classification, due for example to a low score or missing reference spectrum in the database, the 16S rRNA gene sequence remains the gold standard for final identification.

Different platforms from well-established commercial manufacturers are available for MALDI-TOF MS identification of bacteria and yeast; spectral databases are often marketed as part of a proprietary system [5,59].

The critical step to perform this technique directly from positive BC is to remove the blood components to clear any interference and obtain enough concentration of bacteria or yeast to be detected.

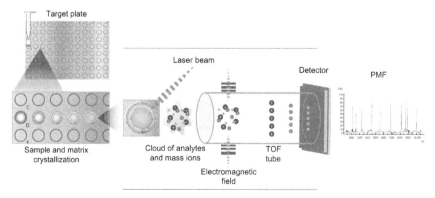

Figure 9.1 MALDI-TOF MS scheme.

Many different studies have been published for this purpose using both in-house and commercial methods. But still being difficult to standardize a protocol, due to the use of different BC bottles, the diversity of pellet obtaining and protein extraction procedures developed (Fig. 9.2) as well as the distinct software for mass analysis used.

A minimum concentration of 10^6 CFU/mL has been considered necessary to obtain high-quality spectra [60,61]. However, some authors report a higher sensitivity of MALDI-TOF MS with sufficient signal-to-noise ratio generated from a sample containing less than 10^4 microorganisms/mL [62–65]. Nevertheless, it is important to have in mind that the inoculum size is difficult to define due to pathogens with poor spectral quality [66].

The common step for each procedure is to process the BCs on the selected instrument as manufacture advice. Bottles are usually incubated for 5 days, and once the instrument indicates that a bottle is positive, a Gram staining is performed and a drop from each positive bottle plastered on the selected agar media and incubated.

The volume of sample from the positive BC used to perform the assay varies, for example, from 200 µL [67] to 1 mL in most of the protocols [65] and up to 8 mL [1], but even largest volumes have been described.

Some of the most performed methods for blood cells removal are the use of gel separator tubes [2,68,69], stepwise centrifugation [4,60,70], lysis-filtration method [71], and the use of different lysis buffer with 5% saponin [65,67], sodium dodecyl sulfate (SDS) [3,20,72], or ammonium

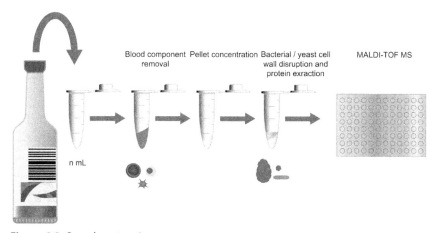

Figure 9.2 Sample extraction.

chloride lysing solution [73] among others. In a recent study Zhou et al., 2017, describes the use of sterile glass beads to get a homogeneous suspension of the pellet to improve bacteria identification especially for Gram-positive bacteria and some Gram-negative bacteria as *Acinetobacter baumannii* or *Klebsiella* spp. as well as the addition of Tween-20 to the pellet, before washing with SDS for yeast identification.

The pellet concentration is usually obtained from cycles of washing and centrifugation with sterile distilled or deionized water and sometimes absolute ethanol.

Regarding the chemical treatment for protein extraction from the pellet, there are plenty of possible combination published, being some of the most common and successful use of trifluoroacetic acid and absolute ethanol [67], trifluoroacetic and acetonitrile [4], combinations of formic acid and acetonitrile [4,74], or just using a formic acid extraction step directly on MALDI-TOF stainless target plate [65].

Another line of investigation to analyze positive BC by MALDI-TOF MS within hours of the same day, it is focused on complete cells instead of protein extraction. This method recovers the microorganisms by a step of centrifugation, in a serum separator tube, followed by the smudge of the sample onto the agar plate. After 2−4 hours, the growing colonies could be studied by MS [68,75].

Currently, the most commonly used commercial systems are VITEK MS (bioMérieux) and Microflex/MALDI Biotyper (Bruker Daltonics, Bremen, Germany), which also has a commercial and standardized lysis kit, Sepsityper kit (Bruker Daltonics). This one has been widely performed and compared with in-house methods, reporting accurate results.

9.3 DATA MANAGEMENT

Results are evaluated according to the manufacturer's standard cut-off values recommended for identification after culture on solid media, establishing the level of identification to species, genus, or no identification, but also using modified cut-off values recommended for methods involving a prior preparation step once obtained an identification score for both bacteria [1,76,77] and yeast [75], increasing the accurate species identification rate [2,22,60].

Several of the published studies have shown that MALDI-TOF MS analysis of samples obtained directly from positive BCs can correctly identify 60%−97% of bacterial and yeast involved in mono-microbial BSI. The level of identification achieved depends on the score level used as cut-off [2−4,60,61,73,75,78−87].

In general, identification of Gram-negative organisms often provides more successful results compared to Gram-positive. This issue has been attributed to suboptimal inoculum size which results in a lack of proteins available for the analysis [58,88,89] but also possibly due to the thick peptidoglycan layer of the cell wall that could affect the extraction process [90]. In case of the lack of identification of *Streptococcus pneumoniae*, some authors explain that it could be conferred to autolysis [65]. Some studies also reported difficulties for the accurate identification of encapsulated bacteria and some species of *Candida*.

The differences between findings of the existing publications could apply to the BC composition, for example, some studies have reported lower identification rates in media containing charcoal [5,67,91] that can inhibit an adequate pellet formation, whereas other studies come across an unalike level of identification while the sample obtained is from the aerobic and the anaerobic BC [1,2,92]. The extraction protocols as well as the software and the available proteomic profile of the database can be involved on the result obtained [2,60,61,67].

Regarding the different sample extraction protocols, more accurate identification results have been reported for Gram-positive microorganisms by incubating the bacterial pellet with 5% saponin [93], 5% SDS [72], or the use of glass beads to get a homogeneous suspension, prior to the protein extraction step with formic acid and acetonitrile [1]. The use of formic acid has also demonstrated better results than trifluoroacetic for protein extraction.

The results obtained from the smudge technique instead of the extraction ones also report high percentages of correct identification for most of the pathogens.

Nevertheless, similar limitations have been found by the numerous publications and are also indicated by manufactures. Some studies have compared the performance of Bruker Biotyper to Vitek MS obtaining similar limitations from both as well [84].

The most representative limitations of the MALDI-TOF MS regarding direct identification from BCs are the misidentification of different species of the *Streptococcus viridans group* (SVG) as *S. pneumoniae* and the identification of poly-microbial infections or mixed cultures.

In relation to misidentification of different species of the SVG, mainly *mitis* group, as *S. pneumoniae* [67,94–96] due to the similar spectra obtained from them, some studies have been published comparing the performance of the two MALDI-TOF MS systems commonly used in clinical microbiology: Bruker Biotyper (Daltonics, German) and Vitek MS (bioMérieux, France); the results achieved by both systems differ on the algorithms used for species identification. Bruker Biotyper (Daltonics, German) generates multiple spectra from a single reference strain and select a consensus one based on the reference spectra; the signals of the sample spectra are compared to the reference assigning a score value. In comparison, Vitek MS (bioMérieux, France) identifies peaks of sample isolates creating bins that are weighted and classified by algorithms that determine the best match for identification [94,97,98].

The analysis of the spectrum of *S. pneumoniae* shows peaks in determined positions that are not usually found in *Streptococcus mitis* and *Streptococcus oralis* isolates [99]. These findings open a path to develop solutions for this particular limitation.

The concern regarding poly-microbial infections or mixed cultures is caused by the difficulty on the identification. The results reported in that case has shown none of the pathogens is identified, just one of the microorganisms is identified, or the different spectra obtained do not score for a reliable classification. However, studies have been performed to find algorithms that may allow resolving this problem by the extraction of every individual peak profile step by step and comparing the spare ones to the reference database 8 or using a specific database based on the BC Gram stain obtained [67].

Also, when more than one morphology are identified by Gram stain microscopy, some authors compare MALDI-TOF MS profiles obtained using a mixed method provided by the software without satisfactory results, so far, but opening another path for analysis enhancement [1].

Despite the fact that further improvements for sample extraction and analytical algorithms for poly-microbial or mixed cultures are required, the identification of more than one microorganism in the same BC provides an important information of pathogens that could not be recognized directly by the initial Gram stain of the sample [3].

9.4 BENEFITS OF USING MALDI-TOF MS TECHNOLOGY AND COST-EFFECTIVENESS

The broad repertoire of available publications regarding the performance of MALDI-TOF MS directly from positive BC for the diagnosis of BSI in clinical microbiology laboratories gets, in general, to define this technique as reliable, accurate, timely, and cost-effective for its purpose.

The cost evaluation of the technique implies to consider several factors that are not easy to settle. However, different studies have been published with this objective, showing positive results regarding the cost-effectiveness of the procedure. Despite the initial purchase of the instrument is considerable, some authors find that the savings entailed could offset the investment from within few years [100−103], reaching percentages of savings above a 55% from the first year [100]. Apart from this initial cost, the evaluations take account of the fixed annual costs of the MALDI-TOF MS, such as the protein standards and instrument maintenance, but also the time of the professionals involved in the review of results and intervention to the therapy is considered.

Savings are reflected on the low costs of reagents and consumables, as well as reduced waste, a decrease of complementary identification techniques in most cases, with the exemption of those pathogens whose identification could not be reliable so far (*S. pneumoniae. . .*). Regarding this issue, the continuous update of the databases would reduce as possible any additional analysis in the future [104]. The studies consider as well for this evaluation the simplicity of the technique but also the fact that no loss of staff is required [102].

Regarding the potential impact that the timely identification has in patient care and hospital costs, the savings are echoed on the appropriate management of the antimicrobial therapy, the related care, and the reduction on the length of stay of the patient by 1 day in most of the cases [101,102,105−107]. Identification is available within hours because the BC becomes positive yielding the patient to receive targeted antibiotic therapy in the first 24 hours after the sample is taken.

Nevertheless, before including MALDI-TOF MS as a routine technique, it should be necessary to consider the annual prevalence of organisms encountered in the laboratory, among other conditions that would allow the laboratory to assume the investment that this technique requires [102]. In this respect, a recent study suggests that an optimization of the analytical processes and procedures surrounding BC systems could be performed before including on the service portfolio this kind of technique [108].

9.5 CONCLUSIONS

Keeping in mind the initial inversion needed to get the MALDI-TOF MS analyzer and despite the lack of a standardized protocol and the limitations of the process, several studies published have demonstrated that this is an easy and cost-effective technique. However, the incorporation of the MALDI-TOF MS as a routine procedure for the identification of pathogens directly from the positive BCs has led to a significant progress, obtaining timely and accurate outcomes which allow clinicians to adjust the antibiotic therapy by reducing the risk of mortality and improving the quality of the management of patient's condition, source control if necessary, and every underlying factor.

ACKNOWLEDGMENTS

I would like to thank the Clinical Microbiology Department team of the Virgen de las Nieves Hospital (Granada, Spain) for offering me the opportunity to collaborate on this project. Moreover, I would like to thank Guillermo Garcia designer of the figures included in this chapter.

REFERENCES

[1] Zhou M, Yang Q, Kudinha T, et al. An improved in-house MALDI-TOF MS protocol for direct cost-effective identification of pathogens from blood cultures. Front Microbiol 2017;8:1824. Available from: https://doi.org/10.3389/fmicb.2017.01824.
[2] Moussaoui W, Jaulhac B, Hoffmann AM, Ludes B, Kostrzewa M, Riegel P, et al. Matrix-assisted laser desorption ionization time-of-flight mass spectrometry identifies 90% of bacteria directly from blood culture vials. Clin Microbiol Infect 2010;16:1631−8.
[3] Rodriguez-Sanchez B, Sanchez-Carrillo C, Ruiz A, Marin M, Cercenado E, Rodriguez-Creixems M, et al. Direct identification of pathogens from positive blood cultures using matrix-assisted laser desorption-ionization time-of-flight mass spectrometry. Clin Microbiol Infect 2014;20:O421−7. Available from: https://doi.org/10.1111/1469-0691.
[4] La Scola B, Raoult D. Direct identification of bacteria in positive blood culture bottles by matrix-assisted laser desorption ionisation time of flight mass spectrometry. PLoS One 2009;4:e8041.

[5] Clark AE, Kaleta EJ, Arora A, Wolk DM. Matrix-assisted laser desorption ionization—time of flight mass spectrometry: a fundamental shift in the routine practice of clinical microbiology. Clin Microbiol Rev 2013;26(3):547−603.

[6] Vlek AL, Bonten MJ, Boel CH. Direct matrix-assisted laser desorption ionization time-of-flight mass spectrometry improves appropriateness of antibiotic treatment of bacteremia. PLoS One 2012;7:e32589.

[7] Seifert H. The clinical importance of microbiological findings in the diagnosis and management of bloodstream infections. Clin Infect Dis 2009;48(suppl 4):S238−45.

[8] Kumar A, Ellis P, Arabi Y, Roberts D, Light B, Parrillo JE, et al. Initiation of inappropriate antimicrobial therapy results in a fivefold reduction of survival in human septic shock. Chest 2009;136(5):1237−48.

[9] Huang AM, Newton D, Kunapuli A, et al. Impact of rapid organism identification via matrix-assisted laser desorption/ionization time-of-flight combined with antimicrobial stewardship team intervention in adult patients with bacteremia and candidemia. Clin Infect Dis 2013;57(9):1237−45. Available from: https://doi.org/10.1093/cid/cit498 PMID: 23899684.

[10] Jamal W, Saleem R, Rotimi VO. Rapid identification of pathogens directly from blood culture bottles by Bruker matrix-assisted laser desorption laser ionization-time of flight mass spectrometry versus routine methods. Diagn Microbiol Infect Dis 2013;76(4):404−8. Available from: https://doi.org/10.1016/j.diagmicrobio 2013.04.013. PMID: 23726652.

[11] Beekmann SE, Diekema DJ, Chapin KC, Doern GV. Effects of rapid detection of bloodstream infections on length of hospitalization and hospital charges. J Clin Microbiol 2003;41(7):3119−25.

[12] van Belkum A, Durand G, Peyret M, Chatellier S, Zambardi G, Schrenzel J, et al. Rapid clinical bacteriology and its future impact. Ann Lab Med 2013;33:14−27.

[13] Banerjee R, Teng CB, Cunningham SA, Ihde SM, Steckelberg JM, Moriarty JP, et al. Randomized trial of rapid multiplex polymerase chain reaction-based blood culture identification and susceptibility testing. Clin Infect Dis 2015;61 (7):1071−80.

[14] Buehler SS, Madison B, Snyder SR, Derzon JH, Cornish NE, Saubolle MA, et al. Effectiveness of practices to increase timeliness of providing targeted therapy for inpatients with bloodstream infections: a laboratory medicine best practices systematic review and meta-analysis. Clin Microbiol Rev 2016;29(1):59−103.

[15] Prowle JR, Echeverri JE, Ligabo EV, et al. Acquired bloodstream infection in the intensive care unit: incidence and attributable mortality. Crit Care 2011;15(2):R100.

[16] Florio W, Morici P, Rizzato C, Barnini S, Tavanti A, et al. Diagnosis of bloodstream infections by mass spectrometry: present and future. Mass Spectrom Open Access 2015;1:106.

[17] Hariu M, Watanabe Y, Oikawa N, Seki M. Usefulness of matrix-assisted laser desorption ionization time-of-flight mass spectrometry to identify pathogens, including polymicrobial samples, directly from blood culture broths. Infect Drug Resist 2017;10:115−20.

[18] Lamy B, Dargère S, Arendrup MC, Parienti J-J, Tattevin P. How to optimize the use of blood cultures for the diagnosis of bloodstream infections? A state-of-the art. Front Microbiol 2016;7:697.

[19] Munson EL, Diekema DJ, Beekmann SE, Chapin KC, Doern GV. Detection and treatment of bloodstream infection: laboratory reporting and antimicrobial management. J Clin Microbiol 2003;41(1):495−7.

[20] Barenfanger J, Graham DR, Kolluri L, Sangwan G, Lawhorn J, et al. Decreased mortality associated with prompt Gram staining of blood cultures. Am J Clin Pathol 2008;130:870−6.

[21] Clerc O, Prod'hom G, Vogne C, Bizzini A, Calandra T, Greub G. Impact of matrix-assisted laser desorption ionization time-of-flight mass spectrometry on the clinical management of patients with gram-negative bacteremia: a prospective observational study. Clin Infect Dis 2013;56(8):1101−7.

[22] Gorton RL, Ramnarain P, Barker K, Stone N, Rattenbury S, McHugh TD, et al. Comparative analysis of Gram's stain, PNA-FISH and Sepsityper with MALDI-TOF MS for the identification of yeast direct from positive blood cultures. Mycoses 2014;57:592−601.

[23] Caspar Y, Garnaud C, Raykova M, Bailly S, Bidart M, Maubon D. Superiority of SDS lysis over saponin lysis for direct bacterial identification from positive blood culture bottle by MALDI-TOF MS. Proteom Clin Appl 2017;11:5−6.

[24] Weinstein MP, et al. Controlled evaluation of BacT/Alert standard aerobic and FAN aerobic blood culture bottles for detection of bacteremia and fungemia. J Clin Microbiol 1995;33(4):978−81.

[25] Wilson ML, Weinstein MP, Mirrett S, Reimer LG, Feldman RJ, et al. Controlled evaluation of BacT/alert standard anaerobic and FAN anaerobic blood culture bottles for the detection of bacteremia and fungemia. J Clin Microbiol 1995;33:2265−70.

[26] McDonald LC, et al. Clinical importance of increased sensitivity of BacT/Alert FAN aerobic and anaerobic blood culture bottles. J Clin Microbiol 1996;34(9):2180−4.

[27] Weinstein MP. Current blood culture methods and systems: clinical concepts, technology, and interpretation of results. Clin Infect Dis 1996;23:40−6.

[28] Akan OA, Yildiz E. Comparison of the effect of delayed entry into 2 different blood culture systems (BACTEC 9240 and BacT/ALERT 3D) on culture positivity. Diagn Microbiol Infect Dis 2006;54(3):193−6.

[29] Sautter RL, et al. Effects of delayed-entry conditions on the recovery and detection of microorganisms from BacT/ALERT and BACTEC blood culture bottles. J Clin Microbiol 2006;44(4):1245−9.

[30] Schwetz I, et al. Delayed processing of blood samples influences time to positivity of blood cultures and results of Gram stain-acridine orange leukocyte cytospin test. J Clin Microbiol 2007;45(8):2691−4.

[31] Piette A, Verschraegen G. Role of coagulase-negative staphylococci in human disease. Vet Microbiol 2009;134(1-2):45−54.

[32] Kirn TJ, Weinstein MP. Update on blood cultures: how to obtain, process, report, and interpret. Clin Microbiol Infect 2013;19:513−20.

[33] Peters RP, van Agtmael MA, Danner SA, Savelkoul PH, Vandenbroucke-Grauls CM. New developments in the diagnosis of bloodstream infections. Lancet Infect Dis 2004;4:751−60.

[34] Dubourg G, Raoult D. Emerging methodologies for pathogen identification in positive blood culture testing. Expert Rev Mol Diagn 2016;16:97−111.

[35] Mitchell SL, Alby K. Performance of microbial identification by MALDI-TOF MS and susceptibility testing by VITEK 2 from positive blood cultures after minimal incubation on solid media. Eur J Clin Microbiol Infect Dis 2017;36:2201.

[36] Gebert S, Siegel D, Wellinghausen N. Rapid detection of pathogens in blood culture bottles by real-time PCR in conjunction with the pre-analytic tool MolYsis. J Infect 2008;57:307−16.

[37] Cattoir V, Gilibert A, Le Glaunec JM, Launay N, Bait-Merabet L, Legrand P. Rapid detection of Pseudomonas aeruginosa from positive blood cultures by quantitative PCR. Ann Clin Microbiol Antimicrob 2010;9:21. Available from: https://doi.org/10.1186/1476-0711-9-21.

[38] Kempf VA, Trebesius K, Autenrieth IB. Fluorescent in situ hybridization allows rapid identification of microorganisms in blood cultures. J Clin Microbiol 2000;38(2):830−8.

[39] Peters RP, Savelkoul PH, Simoons-Smit AM, Danner SA, Vandenbroucke-Grauls CM, van Agtmael MA. Faster identification of pathogens in positive blood cultures by fluorescence in situ hybridization in routine practice. J Clin Microbiol 2006;44:119−23. Available from: https://doi.org/10.1128/JCM.44.1.119-123.2006.

[40] Forrest GN, Roghmann MC, Toombs LS, Johnson JK, Weekes E, Lincalis DP, et al. Peptide nucleic acid fluorescent in situ hybridization for hospital-acquired enterococcal bacteremia: delivering earlier effective antimicrobial therapy. Antimicrob Agents Chemother 2008;52:3558−63.

[41] Calderaro A, Martinelli M, Motta F, Larini S, Arcangeletti MC, Medici MC, et al. Comparison of peptide nucleic acid fluorescence in situ hybridization assays with culture-based matrix-assisted laser desorption/ionization-time of flight mass spectrometry for the identification of bacteria and yeasts from blood cultures and cerebrospinal fluid cultures. Clin Microbiol Infect 2014;20(8):O468−75.

[42] Cleven BE, Palka-Santini M, Gielen J, Meembor S, Kronke M, et al. Identification and characterization of bacterial pathogens causing bloodstream infections by DNA microarray. J Clin Microbiol 2006;44:2389−97.

[43] Wellinghausen N, Wirths B, Franz AR, Karolyi L, Marre R, et al. Algorithm for the identification of bacterial pathogens in positive blood cultures by real-time LightCycler polymerase chain reaction (PCR) with sequence-specific probes. Diagn Microbiol Infect Dis 2004;48:229−41.

[44] Blaschke AJ, Heyrend C, Byington CL, Fisher MA, Barker E, Garrone NF, et al. Rapid identification of pathogens from positive blood cultures by multiplex polymerase chain reaction using the FilmArray system. Diagn Microbiol Infect Dis 2012;74(4):349−55.

[45] Altun O, Almuhayawi M, Ullberg M, Ozenci V. Clinical evaluation of the FilmArray blood culture identification panel in identification of bacteria and yeasts from positive blood culture bottles. J Clin Microbiol 2013;51(12):4130−6.

[46] Vardakas KZ, Anifantaki FI, Trigkidis KK, Falagas ME. Rapid molecular diagnostic tests in patients with bacteremia: evaluation of their impact on decision making and clinical outcomes. Eur J Clin Microbiol Infect Dis 2015;34 (11):2149−60.

[47] Fiori B, D'Inzeo T, Giaquinto A, Menchinelli G, Liotti FM, de Maio F, et al. Optimized use of the MALDI BioTyper system and the FilmArray BCID panel for direct identification of microbial pathogens from positive blood cultures. J Clin Microbiol 2016;54(3):576−84.

[48] Schubert S, Weinert K, Wagner C, Gunzl B, Wieser A, et al. Novel, improved sample preparation for rapid, direct identification from positive blood cultures using matrix-assisted laser desorption/ionization time-of-flight (MALDITOF) mass spectrometry. J Mol Diagn 2011;13:701−6.

[49] Fagerquist CK, Garbus BR, Miller WG, Williams KE, Yee E, Bates AH, et al. Rapid identification of protein biomarkers of Escherichia coli O157:H7 by matrix-assisted laser desorption ionization-time-of-flight-time-of-flight mass spectrometry and top-down proteomics. Anal Chem 2010;82(7):2717−25.

[50] Murray PR. What is new in clinical microbiology—microbial identification by MALDI-TOF mass spectrometry: a paper from the 2011 William Beaumont Hospital Symposium on Molecular Pathology. J Mol Diagn 2012;14(5):419−23.

[51] Larsen MV, Cosentino S, Lukjancenko O, et al. Benchmarking of methods for genomic taxonomy. J Clin Microbiol 2014;52(5):1529−39.

[52] Singhal N, Kumar M, Kanaujia PK, Virdi JS. MALDI-TOF mass spectrometry: an emerging technology for microbial identification and diagnosis. Front Microbiol 2015;6:791. Available from: https://doi.org/10.3389/fmicb.2015.00791.

[53] Tanaka K, Waki H, Ido Y, Akita S, Yoshida Y, Yoshida T, et al. Protein and polymer analyses up to m/z 100 000 by laser ionization time-of-flight mass spectrometry. Rapid Commun Mass Spectrom 1988;2:151−3.

[54] Hillenkamp F, Karas M. Mass spectrometry of peptides and proteins by matrix-assisted ultraviolet laser desorption/ionization. Methods Enzymol 1990;193:280−95.

[55] Albrethsen J. Reproducibility in protein profiling by MALDI-TOF mass spectrometry. Clin Chem 2007;53:852−8.

[56] Dierig A, Frei R, Egli A. The fast route to microbe identification: matrix assisted laser desorption/ionization-time of flight mass spectrometry (MALDI-TOF MS). Pediatr Infect Dis J 2015;34(1):97−9.

[57] Sandrin TR, Goldstein JE, Schumaker S. MALDI TOF MS profiling of bacteria at the strain level: a review. Mass Spectrom Rev 2013;32(3):188−217.

[58] Croxatto A, Prod'hom G, Greub G. Applications of MALDI-TOF mass spectrometry in clinical diagnostic microbiology. FEMS Microbiol Rev 2012;36(2):380−407.

[59] Patel Robin. Matrix-assisted laser desorption ionization−time of flight mass spectrometry in clinical microbiology. Clin Infect Dis 2013;Volume 57(Issue 4):564−72.

[60] Christner M, Rohde H, Wolters M, Sobottka I, Wegscheider K, Aepfelbacher M. Rapid identification of bacteria from positive blood culture bottles by use of matrix-assisted laser desorption-ionization time of flight mass spectrometry fingerprinting. J Clin Microbiol 2010;48(5):1584−91. Available from: https://doi.org/10.1128/JCM.01831-09.

[61] Stevenson LG, Drake SK, Murray PR. Rapid identification of bacteria in positive blood culture broths by matrix-assisted laser desorption ionization-time of flight mass spectrometry. J Clin Microbiol 2010;48(2):444−7. Available from: https://doi.org/10.1128/JCM.01541-09.

[62] Fenselau C, Demirev P. Characterization of intact microorganisms by MALDI mass spectrometry. Mass Spectrom Rev 2001,20:157 71. Available from: https://doi.org/10.1002/mas.10004.

[63] Demirev PA, Fenselau C. Mass spectrometry for rapid characterization of microorganisms. Annu Rev Anal Chem 2008;1:71−93. Available from: https://doi.org/10.1146/annurev.anchem.1.031207.112838.

[64] Hsieh SY, Tseng CL, Lee YS, Kuo AJ, Sun CF, Lin YH, et al. Highly efficient classification and identification of human pathogenic bacteria by MALDI-TOF MS. Mol Cell Proteomics 2008;7:448−56.

[65] Jakovljev A, Bergh K. Development of a rapid and simplified protocol for direct bacterial identification from positive blood cultures by using matrix assisted laser desorption ionization time-of-flight mass spectrometry. BMC Microbiol 2015;15:258. Available from: https://doi.org/10.1186/s12866-015-0594-2.

[66] Friedrichs C, Rodloff AC, Chhatwal GS, Schellenberger W, Eschrich K. Rapid identification of viridans streptococci by mass spectrometric discrimination. J Clin Microbiol 2007;45:2392−7. Available from: https://doi.org/10.1128/JCM.00556-07.

[67] Ferroni A, Suarez S, Beretti J-L, et al. Real-time identification of bacteria and Candida species in positive blood culture broths by matrix-assisted laser desorption ionization-time of flight mass spectrometry. J Clin Microbiol 2010;48(5):1542−8. Available from: https://doi.org/10.1128/JCM.02485-09.

[68] Chen Y, Porter V, Mubareka S, Kotowich L, Simor AE. Rapid identification of bacteria directly from positive blood cultures by use of a serum separator tube, smudge plate preparation, and matrix-assisted laser desorption ionization−time of flight mass spectrometry Burnham C-AD, ed. J Clin Microbiol 2015;53(10):3349−52. Available from: https://doi.org/10.1128/JCM.01493-15.

[69] Wimmer JL, Long SW, Cernoch P, et al. Strategy for rapid identification and antibiotic susceptibility testing of Gram-negative bacteria directly recovered from positive

blood cultures using the Bruker MALDI biotyper and the BD phoenix system. J Clin Microbiol 2012;50(7):2452−4. Available from: https://doi.org/10.1128/JCM.00409-12.

[70] Trenholme GM, Kaplan RL, Karakusis PH, et al. Clinical impact of rapid identification and susceptibility testing of bacterial blood culture isolates. J Clin Microbiol 1989;27(6):1342−5.

[71] Fothergill A, Kasinathan V, Hyman J, Walsh J, Drake T, Wang YF. Rapid identification of bacteria and yeasts from positive-blood-culture bottles by using a lysis-filtration method and matrix-assisted laser desorption ionization−time of flight mass spectrum analysis with the SARAMIS database. J Clin Microbiol 2013;51(3):805−9. Available from: https://doi.org/10.1128/JCM.02326-12.

[72] Hoyos-Mallecot Y, Miranda-Casas C, Cabrera-Alvargonzalez JJ, Gomez-Camarasa C, Perez-Ramirez MD, Navarro-Marı JM. Bacterial identification from blood cultures by a rapid matrix-assisted laser desorption-ionisation time-of-flight mass spectrometry technique. Enferm Infecc Microbiol Clin 2013;31:152−5.

[73] Prod'hom G, Bizzini A, Durussel C, Bille J, Greub G. Matrix-assisted laser desorption ionization-time of flight mass spectrometry for direct bacterial identification from positive blood culture pellets. J Clin Microbiol 2010;48(4):1481−3. Available from: https://doi.org/10.1128/JCM.01780-09.

[74] Freiwald A, Sauer S. Phylogenetic classification and identification of bacteria by mass spectrometry. Nat Protocols 2009;4(5):732−42. Available from: https://doi.org/10.1038/nprot.2009.37.

[75] Idelevich EA, Schule I, Grünastel B, Wüllenweber J, Peters G, Becker K. Rapid identification of microorganisms from positive blood cultures by MALDI-TOF mass spectrometry subsequent to very short-term incubation on solid medium. Clin Microbiol Infect 2014;20:1001−6. Available from: https://doi.org/10.1111/1469-0691.12640.

[76] Nonnemann B, Tvede M, Bjarnsholt T. Identification of pathogenic microorganisms directly from positive blood vials by matrix-assisted laser desorption/ionization time of flight mass spectrometry. APMIS 2013;121:871−7. Available from: https://doi.org/10.1111/apm.12050.

[77] Martinez RM, Bauerle ER, Fang FC. Evaluation of three rapid diagnostic methods for direct identification of microorganisms in positive blood cultures. J Clin Microbiol 2014;52:2521−9. Available from: https://doi.org/10.1128/JCM.00529-14.

[78] Bizzini A, Durussel C, Bille J, Greub G, Prod'hom G. Performance of matrix-assisted laser desorption ionization-time of flight mass spectrometry for identification of bacterial strains routinely isolated in a clinical microbiology laboratory. J Clin Microbiol 2010;48:1549−54.

[79] Seng P, Drancourt M, Gouriet F, et al. Ongoing revolution in bacteriology: routine identification of bacteria by matrix-assisted laser desorption ionization time-of-flight mass spectrometry. Clin Infect Dis 2009;49:543−51.

[80] Ferreira L, Sánchez-Juanes F, Porras-Guerra I, García-García MI, García-Sánchez JE, González-Buitrago JM, et al. Microorganisms direct identification from blood culture by matrixassisted laser desorption/ionization time-of-flight mass spectrometry. Clin Microbiol Infect 2011;17:546−51.

[81] Martiny D, Dediste A, Vandenberg O. Comparison of an in-house method and the commercial Sepsityper™ kit for bacterial identification directly from positive blood culture broths by matrix-assisted laser desorption-ionisation time-of-flight mass spectrometry. Eur J Clin Microbiol Infect Dis 2012;31:2269−81.

[82] Buchan BW, Riebe KM, Ledeboer NA. Comparison of the MALDI Biotyper system using Sepsityper specimen processing to routine microbiological methods for

identification of bacteria from positive blood culture bottles. J Clin Microbiol 2012;50:346—52.

[83] Lagacé-Wiens PRS, Adam HJ, Karlowsky JA, Nichol KA, Pang PF, Guenther J, et al. Identification of blood culture isolates directly from positive blood cultures by use of matrixassisted laser desorption ionization—time of flight mass spectrometry and a commercial extraction system: analysis of performance, cost, and turnaround time. J Clin Microbiol 2012;50:3324—8.

[84] Chen JHK, Ho P-L, Kwan GSW, She KKK, Siu GKH, Cheng VCC, et al. Direct bacterial identification in positive blood cultures by use of two commercial matrix-assisted laser desorption ionization—time of flight mass spectrometry systems. J Clin Microbiol 2013;51:1733—9.

[85] Foster AGW. Rapid identification of microbes in positive blood cultures by use of the Vitek MS matrix-assisted laser desorption ionization-time of flight mass spectrometry system. J Clin Microbiol 2013;51:3717—19.

[86] Loonen AJ, Jansz AR, Stalpers J, Wolffs PF, van den Brule AJ. An evaluation of three processing methods and the effect of reduced culture times for faster direct identification of pathogens from BacT/ ALERT blood cultures by MALDI-TOF MS. Eur J Clin Microbiol Infect Dis 2012;31:1575—83.

[87] Schmidt V, Jarosch A, M€arz P, Sander C, Vacata V, Kalka-Moll W. Rapid identification of bacteria in positive blood culture by matrix-assisted laser desorption ionization time-of-flight mass spectrometry. Eur J Clin Microbiol Infect Dis 2012;31:311—17.

[88] de Cueto M, Ceballos E, Martinez-Martinez L, Perea EJ, Pascual A. Use of positive blood cultures for direct identification and susceptibility testing with the vitek 2 system. J Clin Microbiol 2004;42:3734—8.

[89] Tan TY, Ng LS, Kwang LL. Evaluation of disc susceptibility tests performed directly from positive blood cultures. J Clin Pathol 2008;61:343—6.

[90] Klein S, Zimmermann S, Kohler C, Mischnik A, Alle W, Bode KA. Integration of matrix-assisted laser desorption/ionization time-of-flight mass spectrometry in blood culture diagnostics: a fast and effective approach. J Med Microbiol 2012;61(Pt 3):323—31. Available from: https://doi.org/10.1099/jmm.0.035550-0.

[91] Szabados F, Michels M, Kaase M, Gatermann S. The sensitivity of direct identification from positive BacT/ALERT (bioMerieux) blood culture bottles by matrix-assisted laser desorption ionization time-of-flight mass spectrometry is low. Clin Microbiol Infect 2011;17:192—5.

[92] Almuhayawi M, Altun O, Abdulmajeed AD, Ullberg M, Özenci V. The performance of the four anaerobic blood culture bottles BacT/ALERT-FN, -FN Plus, BACTEC-Plus and -Lytic in detection of anaerobic bacteria and identification by direct MALDI-TOF MS Woo PC, ed. PLoS One 2015;10(11):e0142398. Available from: https://doi.org/10.1371/journal.pone.0142398.

[93] Meex C, Neuville F, Descy J, et al. Direct identification of bacteria from BacT/ ALERT anaerobic positive blood cultures by MALDI-TOF MS: MALDI Sepsityper kit versus an in-house saponin method for bacterial extraction. J Med Microbiol 2012;61:1511—16.

[94] Zhou M, Yang Q, Kudinha T, Zhang L, Xiao M, Kong F, et al. Using matrix-assisted laser desorption ionization-time of flight (MALDI-TOF) complemented with selected 16S rRNA and gyrB genes sequencing to practically identify clinical important viridans group streptococci (VGS). Front Microbiol 2016;7:1328. Available from: https://doi.org/10.3389/fmicb.2016.01328.

[95] Ikryannikova LN, Lapin KN, Malakhova MV, Filimonova AV, Ilina EN, Dubovickaya VA, et al. Misidentification of alpha-hemolytic streptococci by routine tests in clinical practice. Infect Genet Evol 2011;11:1709—15. Available from: https://doi.org/10.1016/j.meegid.2011.07.010.

[96] Davies AP, Reid M, Hadfield SJ, Johnston S, Mikhail J, Harris LG, et al. Identification of clinical isolates of alpha-hemolytic streptococci by 16S rRNA gene sequencing, matrix-assisted laser desorption ionization-time of flight mass spectrometry using MALDI Biotyper, and conventional phenotypic methods: a comparison. J Clin Microbiol 2012;50:4087−90. Available from: https://doi.org/10.1128/JCM.02387-12.

[97] Kärpänoja P, Harju I, Rantakokko-Jalava K, Haanperä M, Sarkkinen H. Evaluation of two matrix-assisted laser desorption ionization-time of flight mass spectrometry systems for identification of viridans group streptococci. Eur J Clin Microbiol Infect Dis 2014;33:779−88. Available from: https://doi.org/10.1007/s10096-013-2012-8.

[98] Angeletti S, Dicuonzo G, Avola A, Crea F, Dedej E, Vailati F, et al. Viridans group Streptococci clinical isolates: MALDI-TOF mass spectrometry versus gene sequence-based identification. PLoS One 2015;10:e0120502. Available from: https://doi.org/10.1371/journal.pone.0120502.

[99] Werno AM, Christner M, Anderson TP, Murdoch DR. Differentiation of Streptococcus pneumoniae from nonpneumococcal streptococci of the Streptococcus mitis group by matrix-assisted laser desorption ionization-time of flight mass spectrometry. J Clin Microbiol 2012;50:2863−7.

[100] Neville SA, et al. Utility of matrix-assisted laser desorption ionization-time of flight mass spectrometry following introduction for routine laboratory bacterial identification. J Clin Microbiol 2011;49:2980−4.

[101] Tan K, Ellis B, Lee R, Stamper P, Zhang S, Carroll K. Prospective evaluation of a matrix-assisted laser desorption ionization−time of flight mass spectrometry system in a hospital clinical microbiology laboratory for identification of bacteria and yeasts: a bench-by-bench study for assessing the impact on time to identification and costeffectiveness. J Clin Microbiol 2012;50:3301−8. Available from: https://doi.org/10.1128/JCM.01405-12.

[102] Patel TS, Kaakeh R, Nagel JL, Newton DW, Stevenson JG. Cost analysis of implementing matrix-assisted laser desorption ionization−time of flight mass spectrometry plus real-time antimicrobial stewardship intervention for bloodstream infections. J Clin Microbiol 2017;55:60−7. Available from: https://doi.org/10.1128/JCM.01452-16.

[103] Gaillot O, et al. Cost-effectiveness of switch to matrix-assisted laser desorption ionization−time of flight mass spectrometry for routine bacterial identification. J Clin Microbiol 2011;49:4412.

[104] Cherkaoui A, et al. Comparison of two matrix-assisted laser desorption ionization−time of flight mass spectrometry methods with conventional phenotypic identification for routine identification of bacteria to the species level. J Clin Microbiol 2010;48:1169−75.

[105] Perez KK, Olsen RJ, Musick WL, Cernoch PL, Davis JR, Land GA, et al. Integrating rapid pathogen identification and antimicrobial stewardship significantly decreases hospital costs. Arch Pathol Lab Med 2013;137:1247−54.

[106] McElvania TeKippe E. The added cost of rapid diagnostic testing and active antimicrobial stewardship: is it worth it? J Clin Microbiol 2017;55:20−3. Available from: https://doi.org/10.1128/JCM.02061-16.

[107] Perez KK, Olsen RJ, Musick WL, Cernoch PL, Davis JR, Peterson LE, et al. Integrating rapid diagnostics and antimicrobial stewardship improves outcomes in patients with antibiotic-resistant Gram negative bacteremia. J Infect 2014;69:216−25.

[108] Banerjee Ritu, Özenci Volkan, Patel Robin. individualized approaches are needed for optimized blood cultures. Clin Infect Dis 2016;63(10):1332−9. Available from: https://doi.org/10.1093/cid/ciw573.

FURTHER READING

French K, Evans J, Tanner H, Gossain S, Hussain A. The clinical impact of rapid, direct MALDI-ToF identification of bacteria from positive blood cultures Anjum M, ed. PLoS One 2016;11(12):e0169332. Available from: https://doi.org/10.1371/journal. pone.0169332.

Wessels E, Schelfaut JJ, Bernards AT, Claas EC. Evaluation of several biochemical and molecular techniques for identification of *Streptococcus pneumoniae* and *Streptococcus pseudopneumoniae* and their detection in respiratory samples. J Clin Microbiol 2012;50:1171−7. Available from: https://doi.org/10.1128/JCM.06609-11.

CHAPTER TEN

Use of MALDI-TOF Techniques in the Diagnosis of Urinary Tract Pathogens

Fernando Sánchez-Juanes[1], Alicia Inés García Señán[2,3], Sara Hernández Egido[3], María Siller Ruiz[3], José Manuel González Buitrago[1,4,5] and Juan Luis Muñoz Bellido[2,3,6]

[1]Instituto de Investigación Biomédica de Salamanca (IBSAL), Universidad de Salamanca, CSIC, Complejo Asistencial Universitario de Salamanca, Salamanca, Spain
[2]Servicio de Microbiología y Parasitología, Complejo Asistencial Universitario de Salamanca, Salamanca, Spain
[3]Research Group on Clinical Microbiology and Parasitology and Antimicrobial Resistance (IIMD-16), Instituto de Investigación Biomédica de Salamanca (IBSAL), Universidad de Salamanca, CSIC, Complejo Asistencial Universitario de Salamanca, Salamanca, Spain
[4]Servicio de Análisis Clínicos, Complejo Asistencial Universitario de Salamanca, Salamanca, Spain
[5]Departamento de Bioquímica y Biología Molecular, Universidad de Salamanca, Salamanca, Spain
[6]Departamento de Ciencias Biomédicas y del Diagnóstico, Universidad de Salamanca, Salamanca, Spain

10.1 INTRODUCTION: EPIDEMIOLOGY AND IMPORTANCE OF UTIS

The term urinary tract infections (UTIs) encompass a wide group of clinical entities different in symptoms, treatment, and potential severity: asymptomatic bacteriuria, acute uncomplicated cystitis, recurrent cystitis, catheter-associated asymptomatic bacteriuria and UTI, pyelonephritis, and prostatitis. UTIs are the second cause of medical consultation in the primary healthcare setting, after respiratory tract infections, and affect mainly sexually active young women. In the hospital settings, UTI is the most frequent infectious complication. For anatomical reasons, women are more likely to develop UTIs. Of note, 50%−60% of women will have at least one UTI episode in their lifetime, and this accounts for an important morbidity and healthcare costs, estimated around US $3.5 billion/year in the United States [1−3]. The risk for developing an UTI increases with sexual intercourse, pregnancy, and age.

Most UTIs are uncomplicated, mild to moderate infections, diagnosed in the community settings. Uncomplicated UTIs are defined as those affecting lower urinary tract in the absence of an underlying condition, functional or structural abnormalities of the genitourinary tract, or recent urologic instrumentation. Nevertheless, complicated UTIs are not infrequent and are associated to several factors such as congenital or acquired urinary tract anatomical abnormalities, obstruction (urinary stones, prostatic hyperplasia...), indwelling catheters, urinary tract surgery, etc. In male, any UTI is considered complicated.

Complicated UTIs are more difficult to treat (it normally takes longer medication regimen) and deserve further urological studies frequently. In addition, recurrent (posttherapy) episodes (relapse or reinfection) are more common than in noncomplicated UTIs. Hospital-acquired UTIs are most frequently associated to urethral catheterization (98% in ICUs [4]) and are currently the leading cause for healthcare associated bacteremia. Approximately, 20% of hospital-acquired bloodstream infections arise from the urinary tract, and the mortality associated with this condition is about 10% [5]. In such situations, an accurate and early antimicrobial therapy is essential as well as removing the urethral catheter when possible.

Escherichia coli is isolated in >90% of uncomplicated cystitis and pyelonephritis. Other Gram-negative bacilli belonging to the *Enterobacteriaceae* family (*Proteus mirabilis, Klebsiella pneumoniae*...) are less common [1,3,6]. *Staphylococcus saprophyticus* causes 5%−10% of uncomplicated UTIs. Fungal UTIs are less frequent than bacterial UTIs, but patients with indwelling catheters, diabetes, immunosuppression, and recent antibiotic use have increased risk of fungal infection [7,8]. The most common microorganism involved in ICU-acquired UTIs is also *E. coli*, followed by *Enterococcus* spp., *Pseudomonas aeruginosa*, and *Klebsiella* spp. [9]. The antibiotic resistance rates among antibiotics commonly used for the treatment of UTIs is increasing. According to a multicenter study carried out in Spain, fluoroquinolone and trimethoprim/sulfamethoxazole resistance rates were 12% and 34%, respectively [10].

The high frequency, potential risk of complications, and healthcare-related costs associated to UTIs make the development of technologies important, which is able to expedite diagnosis, identify pathogens, and enable etiologic treatment.

10.2 CONVENTIONAL DIAGNOSIS OF UTIS

As a rule, screening for asymptomatic bacteriuria is not recommended in man and nonpregnant women. Asymptomatic bacteriuria in pregnancy is associated with symptomatic UTI, preterm labor, and low birthweight, and treatment is indicated because it reduces the risk of these complications. Screening is also indicated for urinary tract instrumentation that can result in mucosal bleeding, because detection and treatment of bacteriuria in these patients reduce the risk of bacteremia.

When the patient is a woman, clinical symptoms of UTI (dysuria, urgency, frequency) are evident, and nothing suggests alternative diagnosis or complications, UTIs can be treated without further testing, so screening methods and urine cultures are not necessary. In any other patients, screening and culture are necessary for diagnosis and treatment. Fast laboratory methods, such as a urine dipstick for leukocyte esterase, Gram staining, and other screening tests based on different technologies, such as urine flow cytometry, have been proposed, alone or combined, for rapid diagnosis of UTIs.

Urine culture remains the gold standard method for UTI diagnosis. Urine culture has a high sensitivity and specificity, allows a specific bacterial count and the detection of mixed culture and contaminations, and is necessary for antibiotic susceptibility tests. Nevertheless, it is time-consuming, may lead to significant workload increase, and requires 24–48 hours for definitive results [11]. Because up to 60% of these specimens yield no growth [12,13], different screening techniques have been proposed to select which samples deserve being processed. The dipstick chemical test is based on strips that allow a semiquantitative study of nitrite (a common product of urinary tract pathogens), leukocyte esterase, proteins, and blood. This method has a high positive predictive value when both nitrite and leukocyte esterase are positive [14], but can be frequently negative in UTIs with a low bacterial count [15]. The microscopic examination of urine sediment and count of cells, particles, and microorganisms is subject to interpretation errors and interindividual variability [16]. Although these older screening methods are still used, they are subjective and time-consuming or have a poor sensitivity and/or specificity. Flow cytometry-based detection and quantification of leukocytes and bacteria has been proposed for UTI screening [17]. The Gram staining has also been used as a screening method, although it can be as time-consuming as the urine culture.

10.3 MATRIX-ASSISTED LASER DESORPTION IONIZATION—TIME-OF-FLIGHT MASS SPECTROMETRY AS A UTI SCREENING TOOL

Matrix-assisted laser desorption ionization—time-of-flight mass spectrometry (MALDI-TOF MS) has been shown as a very reliable method for identification of urinary tract pathogens, but also as an excellent method for identification of microorganisms directly from samples. In fact, direct identification might be considered as an improved screening method, because it allows, with some limitations, both differentiation between positive and negative samples and bacterial identification. The first approach to direct identification of bacterial urinary tract pathogens from urine samples, published in 2010, already suggested this possibility [18]. In this study, 260 samples from inpatients and outpatients with symptoms suggesting UTI were studied by conventional methods, including flow cytometry screening, and by MALDI-TOF MS. Twenty samples were considered false positive (were reported as positive by the flow cytometry screening, but were negative in culture), and all the 20 samples were negative in MALDI-TOF MS, thus specificity of MALDI-TOF MS seems higher. Nevertheless, 14 samples neatly positive, with colony counts $>10^5$ CFU/mL and 14 samples with colony counts $<10^5$ CFU/mL, thus potentially positive in specific situations, were reported as "no peaks detected" by MALDI-TOF MS. Because in screening tests sensitivity shall prevail over specificity, these results suggests that MALDI-TOF MS might be useful as a part of a screening strategy alternative to conventional procedures, but probably not as a screening method itself. Burillo et al. [19] also proposed a method in which they combine Gram staining and MALDI-TOF MS, reporting results in less than 1 hour. The use of both techniques anticipated the culture result in 82.7% of cases, gave information with minor errors (information partially anticipative of the culture result) in 13.4% and provided information with major errors (incorrect results, potentially leading to inappropriate changes in antimicrobial therapy) in 3.9%. Thus, with independence of the accuracy of identification, the combination of flow cytometry or Gram stain with MALDI-TOF MS, can report very accurate results, in terms of positivity or negativity, in a very short time. Additionally, the accuracy of identification obtained allows a more adjusted treatment.

10.4 IDENTIFYING UTI PATHOGENS BY MALDI-TOF MS

Introduction of MALDI-TOF MS in microbiology laboratories has been surprisingly fast and easy, due to its simplicity and reliability, especially for the identification of microorganisms, which is its main use. A number of studies have shown that it is a diagnostic technology extremely reliable for identifying the most frequent human pathogens [20], but also for microorganisms of infrequent isolation whose identification can be complicated with conventional methods [21].

In 2010, our group published a study in which we observed a high level of correlation between MALDI-TOF MS and conventional identification methods. Working on colonies grown on agar plates, the correlation was 100%, to the species level, in Gram positives, and 97.7% to the genus level and 87.7% to the species level for Gram negatives [22].

As for the microorganisms most usually causing UTI, most of them are included in these good results. We can find some limitations, as it is the case of some yeast [23] and some Gram positives. These deficiencies are thought to be more related to protein extraction methods that may not be fully adequate for some microorganisms, or to insufficiencies of the databases, and not to a lack of specificity of the MS [24].

10.4.1 Gram-negative bacteria

Most bacteria producing UTIs are, in fact, Gram-negative bacilli (*E. coli,* *Klebsiella* spp., *Proteus* spp. ...). All these genera have excellent results with MS.

Some genera, such as *Enterobacter* and *Citrobacter*, may show some inaccuracy in the identification to the species level. In the case of *Enterobacter* spp. the first studies only reached 80% sensitivity, [25] data that have subsequently been clearly improved. The results of *Citrobacter* spp. have improved substantially with agreement around 90%—95%. However, some species, due to their close relationship, still are not differentiated accurately [26]. Nevertheless, in whole, the accuracy of MALDI-TOF MS identifications is much higher than previous conventional systems. In some cases, the development and diffusion of MALDI-TOF MS has revealed the real importance of some genera, as it has been shown for the genus *Raoultella*, undervalued because isolates belonging to this genus were usually identified as *Klebsiella* (usually *Klebsiella oxytoca*) [27].

Concerning non-fermenter Gram-negative bacilli, both *Pseudomonas* spp. and *Acinetobacter* show a high reliability at the genus level, close to 100%, but the reliability to the species level is more irregular in different studies [28,29]. *Stenotrophomonas maltophilia* identification also shows a very high reliability [30].

Isolation of *Salmonella* spp. from UTIs is unusual, but it happens occasionally. In the first publications concerning MALDI-TOF bacterial identification, the identification of *Salmonella* at the genus level is reliable, the same does not occur at the species level. However, there have been publications since 2004 that suggest the existence of specific peaks that could allow the identification of serovars with greater reliability [31]. Because there seem to be differential elements, an extension of the reference database with a greater number of spectra of, at least, the more frequent serovars, would help us to solve this limitation, improving the results at the species level.

10.4.2 Gram-positive bacteria

Due to Gram-positive wall composition, with a greater amount of peptidoglycan, the results obtained with MS are not as good as the excellent results obtained with Gram-negative bacteria [22,32]. For some species, additional procedures may be necessary, such as the use of lysostaphin, lysozyme, and other molecules that help break the wall [33].

In the case of *Staphylococcus* genus, comparative studies between the MALDI-TOF technology and phenotypic and molecular studies show excellent correlation for *Staphylococcus aureus*, which can act as an occasional UTI pathogen. A worse correlation to the species level has been shown for coagulase nonproducer, mainly in the first studies [34]. However, the improvement of the technology and mainly the improvement of databases allow now the identification of 97.7%—100% to the species level, compared to 77.5%—76% of other automated systems. [35,36]. This improvement is important for the accurate identification of *S. saprophyticus*, an important UTI pathogen, especially in young, sexually active women.

As a whole, the correct identification of *Streptococcus* through MALDI-TOF MS with a high degree of reliability has always shown significant problems, such as the differentiation between different species of *viridans* streptococci [37]. Nevertheless, MALDI-TOF MS is fully reliable for the identification of those streptococci most usually involved in UTI, such as

Streptococcus pyogenes [38] and *Streptococcus agalactiae.* Concerning *S. agalactiae*, identification directly from a selective enrichment broth has been shown reliable in samples from pregnant women [39].

Reliability of MALDI-TOF MS for identifying both *Enterococcus faecium and Enterococcus faecalis* has been shown very high [40]. Fang et al. [41] concluded in their study that, in the case of enterococci, MALDI-TOF MS is an identification technique at least as reliable as conventional identification, and obtaining results is significantly faster.

Another genus including urinary tract pathogens, for which MALDI-TOF MS has been relevant, has been *Corynebacterium*. UTIs caused by corynebacteria had classically two problems: the slow growth, which meant that many ICUs were not diagnosed if the plates were kept in incubation for only 24 hours, and the low accuracy of identification of some species with conventional, biochemical methods. MALDI-TOF MS has been shown very reliable for the identification of *Corynebacterium urealyticum* and other corynebacteria eventually involved in UTIs [42].

We cannot forget other microorganisms that, because of its slow growth or its difficulty to grow in conventional media and the difficult identification by conventional methods, have been underdiagnosed for years. The introduction of MALDI-TOF MS in clinical microbiology laboratory has increased the number of reports of these microorganisms, as it is the case of *Actinotignum schaalii* [43].

10.4.3 Mycology

Yeasts identification with MALDI-TOF MS shows a high reliability, similar to bacteria. The protein extraction method, the growth conditions, and the incubation time have been shown to have significant influence on the results obtained [44]. Recent studies show a sensitivity of 82.7%–87.2% [45]. Identification failures are usually associated mainly to the absence of specific species in the databases, more than to technical problems for processing and identifying fungal proteins [36].

10.5 DIRECT DIAGNOSIS OF UTIS BY MALDI-TOF MS

Urine culture remains the gold standard for the diagnosis of UTI. Nevertheless, this procedure requires at least 18 hours for the detection of bacterial growth in the culture plates by the usual microbiological techniques. This way, the diagnosis confirmation is delayed at least 24 hours since the sample is received in the laboratory, and the identification of the specific pathogen and the antibiogram is delayed for at least 48 hours.

Thus, methods allowing UTI diagnosis and microorganisms identification in a shorter time would mean a significant improvement. The application of MALDI-TOF MS directly on urine samples has aroused great interest in the last years. In theory, urine is a good sample for direct application of MALDI-TOF MS, because negative samples are sterile or have a bacterial count too low for being detected by this method, while positive samples have a very high bacterial load (usually $>10^5$ CFU/mL) of one only bacterial species. Moreover, the sample volume is not limiting. This way, MALDI-TOF MS would work at the same time as screening and as identification method, reporting pathogens identification in minutes.

We published in 2010, for the first time, a study in which we directly identified by MALDI-TOF MS microorganisms in urine samples [18]. The method included a selection of positive urines by an automated screening device based on flow cytometry (UF-1000i; BioMérieux, Marcy lÉtoile, France). These samples were cultured both in blood and MacConkey agar plates and, after 18 hours of incubation at 37°C, they were identified by conventional methods

In parallel, the same urine samples were processed for MALDI-TOF MS identification. The procedure uses 6 mL of urine that is centrifuged at low-revolution setting (2000 ×g) to remove leukocytes and then at high revolutions (15,500 ×g) to collect bacteria. The pellet is washed and then applied directly to the MALDI-TOF plate. The result with the identification of the microorganism is obtained in minutes. When no results were obtained from direct method, an extraction method was applied. Briefly, bacteria were resuspended in 300 μL of water and then 900 μL of ethanol were added. Mixture was centrifuged (15,000 ×g, 2 minutes) and supernatant discarded. Formic acid (70% [vol/vol]) and acetonitrile were added, and the mixture was centrifuged again to extract proteins. Supernatant (1 μL) was spotted twice on the MALDI target plate. Both direct and extraction spots were overlaid with matrix solution (α-cyano-4-hydroxycinnamic acid) and processed under usual MALDI-TOF routines for bacterial identification (MALDI Biotyper; Bruker Daltonics, Bremen, Germany).

The results obtained in this study indicated that MALDI-TOF MS allows bacterial identification directly from infected urine in a short time, with high accuracy, especially with Gram-negative bacteria with high bacterial count ($>10^5$ CFU/mL). MALDI-TOF MS identified correctly the microorganism directly from urine in 91.8% of the cases to the species level and in 92.7% to the genus level. Failures were due mainly to bacterial counts lower than 10^5 CFU/mL. With higher colony counts, the correct identification rate was as high as 97.6%.

Thus, using this threshold of 10^5 CFU/mL, there is almost complete assurance that the identification correlates with the UTI and, according to the results, allows diagnosing the etiology of more than 90% of UTI in a few minutes. This publication opened a new track that has generated more related publications that take the basis of the method proposed by Ferreira et al. with minor modifications in order to improve percentages of identification or polish the protocol to make it easier. All these studies, in whole, corroborate the results obtained in our first study [19,46].

Other variations of this protocol have been proposed; therefore, numerous publications have been generated. Three years later, Wang et al. [47] analyzed a higher number of samples and they included urines with two different microorganisms in their study. Similar rates of correct identifications were obtained, and they corroborated the importance of colony counts.

In order to improve MALDI-TOF MS direct identification, we added a sodium dodecyl sulfate (SDS) pretreatment to our protocol [48]. Some 7%–8% of samples were not correctly identified despite having a high bacterial count and being a species included in the database. We proposed this study based on the possibility that this was due to a preferably intra-leukocytic presence of the microorganisms, and that identification failures reported in culture-positive samples might be associated with the removal of intraleukocytic microorganisms in these first steps of processing. In this sense, a pretreatment with SDS was included, because this compound can lysate cells and then release microorganisms, thereby increasing method sensitivity. The new method allowed a correct identification of 46% of samples that could not be identified with the first described method, reaching figures around 96%–97.5% of correct identifications. As happened with the standard method, the results are better for Gram-negative microorganisms. Nevertheless, 32% of urine samples harboring Gram-positive bacteria that had not been identified by the standard method were identified with the SDS pretreatment.

Burillo et al. [19] proposed a Gram staining as fast screening method, followed by MALDI-TOF MS analysis after Gram interpretation. They developed a sequential algorithm that combined both results; this resulted in similar rates of correct identification as compared with previously described methods [18,48].

March Roselló et al. proposed in 2015 [49] a new method of direct identification based on two major concepts: a differential centrifugation protocol and new validation criteria. These criteria can be divided into two groups depending on the score and the index of different readings, and another one based only in the score of the first microorganism in the list. They suggested a new interpretation of the score values proposed by the manufacturer: if score values greater than 1.4 for the same first micro-organism on the list were obtained in at least two of four readings, this would be enough to get a correct identification. Eventually if they got similar conclusions that had been suggested before, high bacteria counts are needed to get a good direct identification.

Recently, Íñigo et al. [50] evaluated again the capacity of MALDI-TOF MS to identify microorganisms directly from urine samples. They also analyzed the predictive value of automated analyzers (a flow cytometer (Sysmex UF-1000i) and an automatic sediment analyzer with microscopy (SediMax)). The method used for direct study in MALDI-TOF MS was similar to that described by Ferreira et al. [18]. Identification with MALDI-TOF was correct in 74.3% of cases and was 86.10% for Gram negatives. They also established the best bacteriuria cut-offs for UF-1000i and SediMax. They concluded that the combination of a urine screening method and MALDI-TOF MS provides a reliable identification from urine samples, especially when Gram-negative bacteria are the etiological agent.

10.6 STRATEGIES FOR MALDI-TOF MS ANTIBIOTICS SUSCEPTIBILITY TESTING IN UTIS

The available studies suggest, on the whole, that the direct application of MALDI-TOF MS to urine samples offers good results, mainly with Gram-negative bacteria, and especially if it is applied after an automated screening. The biggest problem that can be put in this moment is that it has not been possible to develop a system that allows guiding the treatment with a similar speed. Zboromyrska et al. [46] proposed the addition of an antimicrobial susceptibility testing (AST). This AST would be performed directly on the remaining pellet after direct bacterial ID by MALDI-TOF-MS. An equivalent 0.5 McFarland suspension was prepared in sterile saline with the remaining pellet, and this suspension was plated on Mueller Hinton agar to perform a disc diffusion method. The results of the direct susceptibility testing correlated with conventional AST in all the monomicrobial UTIs (most of them Gram negatives). The turn-around time of the protocol for UTI diagnosis and microbial identification was 1 hour, and for AST was 18−24 hours.

In whole, obtaining through proteomic techniques, susceptibility results similar to those provided by conventional bacteriological techniques, with a speed and reliability similar to those achieved in identification, are still a major challenge. Methods have been developed to detect some specific resistance mechanisms (ESBLs, carbapenemases, *vanB*) [51−52]. These strategies are not likely to provide global susceptibility profiles comparable to conventional antibiograms. New methodologies have been suggested based in a short incubation of the same pellet used for MALDI-TOF MS−based identification with specific concentrations of antibiotics. These methods has been shown useful for the identification of methicillin-resistant *S. aureus*, in the detection of resistance to β-lactams, aminoglycosides, and fluoroquinolones, of carbapenemase-producing *Klebsiella* spp., and more recently have been tested with other microorganisms against several antibiotics, with good results [53−55]. No papers have still been published on these method applied directly to positive urine samples, but the good results obtained with blood cultures suggest that it might be a good alternative to obtain reliable susceptibility results in 2−3 hours since the screening positivity.

REFERENCES

[1] Foxman B. The epidemiology of urinary tract infection. Nat Rev Urol 2010;7:653−60.
[2] Griebling TL. Urologic diseases in America project: trends in resource use for urinary tract infections in women. J Urol 2005;173:1281−7.
[3] Davenport M, Mach KE, Shortliffe LMD, Banaei N, Wang TH, Liao JC. New and developing diagnostic technologies for urinary tract infections. Nat Rev Urol 2017;14:296−310.
[4] Nicolle LE. Urinary tract infection. Crit Care Clin 2013;29:699−715.
[5] Wagenlehner FM, et al. Diagnosis and management for urosepsis. Int J Urol 2013;20:963−70.
[6] Wilson ML, Gaido L. Laboratory diagnosis of urinary tract infections in adult patients. Clin Infect Dis 2004;38:1150−8.
[7] Kauffman CA. Diagnosis and management of fungal urinary tract infection. Infect Dis Clin North Am 2014;28:61−74.
[8] Sobel JD, Fisher JF, Kauffman CA, Newman CA. Candida urinary tract infections—epidemiology. Clin Infect Dis 2011;52(Suppl 6):S433−6.
[9] Healthcare-associated infections acquired in intensive care units. Annual Epidemiological Report 2016 (2014 data). ECDC.

[10] Palou J, Pigrau C, Molina I, Ledesma JM, Angulo J. Etiología y sensibilidad de los uropatógenos identificados en infecciones urinarias bajas no complicadas de la mujer (Estudio ARESC): implicaciones en la terapia empírica. Med Clin 2011;136:1−7.

[11] Davies EM, Lewis DA. Bacteriology of urine. In: Hawkey P, Lewis DA, editors. Medical bacteriology. Oxford University Press; 2004. p. 1−25.

[12] Broeren MAC, Bahçeci S, Vader HL, Arents NLA. Screening for urinary tract infection with the sysmex UF-1000i urine flow cytometer. J Clin Microbiol 2011;49:1025−9.

[13] Jolkkonen S, Paattiniemi EL, Kärpänoja P, Sarkkinen H. Screening of urine samples by flow cytometry reduces the need for culture. J Clin Microbiol 2010;48:3117−21.

[14] Berger RE. The urine dipstick test useful to rule out infections. A meta-analysis of the accuracy. J Urol 2005;174:941−2.

[15] Waisman Y, Zerem E, Amir L, Mimouni M. The validity of the uriscreen test for early detection of urinary tract infection in children. Pediatrics 1999;104:e41.

[16] Wiwanitkit V, Udomsantisuk N, Boonchalermvichian C. Diagnostic value and cost utility analysis for urine Gram stain and urine microscopic examination as screening tests for urinary tract infection. Urol Res 2005;33:220−2.

[17] Mejuto P, Luengo M, Díaz-Gigante J. Automated flow cytometry: an alternative to urine culture in a routine clinical microbiology laboratory? Int J Microbiol 2017;8532736.

[18] Ferreira L, Sánchez-Juanes F, González-Avila M, Cembrero-Fuciños D, Herrero-Hernández A, González-Buitrago JM, et al. Direct identification of urinary tract pathogens from urine samples by matrix-assisted laser desorption ionization-time of flight mass spectrometry. J Clin Microbiol 2010;48:2110−15.

[19] Burillo A, Rodríguez-Sánchez B, Ramiro A, Cercenado E, Rodríguez-Créixems M, Bouza E. Gram-stain plus MALDI-TOF MS (matrix-assisted laser desorption ionization-time of flight mass spectrometry) for a rapid diagnosis of urinary tract infection. PLoS One 2014;22:e86915.

[20] Seng P, Drancourt M, Gouriet F, La Scola B, Fournier PE, Rolain JM, et al. Ongoing revolution in bacteriology: routine identification of bacteria by matrix-assisted laser desorption ionization time-of-flight mass spectrometry. Clin Infect Dis 2009;49:543−51.

[21] Porras MI, Cañueto J, Ferreira L, García MI. Human dermatophilosis. First description in Spain and diagnosis by matrix-assisted laser desorption ionization time-of-flight mass spectrometry (MALDI-TOF). Enferm Infecc Microbiol Clin 2010;28:747−8.

[22] Ferreira L, Vega S, Sánchez-Juanes F, González M, Herrero A, Muñiz MC, et al. Identifying bacteria using a matrix-assisted laser desorption ionization time-of-flight (MALDI-TOF) mass spectrometer. Comparison with routine methods used in clinical microbiology laboratories. Enferm Infecc Microbiol Clin 2010;28:492−7.

[23] Galán F, García-Agudo L, Guerrero I, Marín P, García-Tapia A, García-Martos P, et al. Evaluación de la espectrometría de masas en la identificación de levaduras de interés clínico. Enferm Infecc Microbiol Clin 2015;33:273−8.

[24] Muñoz Bellido JL, González Buitrago JM. Espectrometría de masas MALDI-TOF en microbiología clínica. Situación actual y perspectivas futuras. Enferm Infecc Microbiol Clin 2015;33:369−71.

[25] Pavlovic M, Konrad R, Iwobi AN, Sing A, Busch U, Huber I. A dual approach employing MALDI-TOF MS and real-time PCR for fast species identification within the *Enterobacter cloacae* complex. FEMS Microbiol Lett 2012;328:46−53.

[26] Febbraro F, Rodio DM, Puggioni G, Antonelli G, Pietropaolo V, Trancassini M. MALDI-TOF MS versus VITEK®2: comparison of systems for the identification of microorganisms responsible for bacteremia. Curr Microbiol 2016;73:843−50.

[27] Ponce-Alonso M, Rodríguez-Rojas L, Del Campo R, Cantón R, Morosini MI. Comparison of different methods for identification of species of the genus *Raoultella*: report of 11 cases of *Raoultella* causing bacteraemia and literature review. Clin Microbiol Infect 2016;22:252−7.

[28] Mulet M, Gomila M, Scotta C, Sanchez D, Lalucat J, Garcia-Valdes E. Concordance between whole-cell matrix-assisted laser-desorption/ionization time-of-flight mass spectrometry and multilocus sequence analysis approaches in species discrimination within the genus *Pseudomonas*. Syst Appl Microbiol 2012;35:455−64.

[29] Hsueh P-R, Kuo L-C, Chang T-C, Lee T-F, Teng S-H, Chuang Y-C, et al. Evaluation of the Bruker Biotyper matrix-assisted laser desorption ionization-time of flight mass spectrometry system for identification of blood isolates of *Acinetobacter* species. J Clin Microbiol 2014;52:3095−100.

[30] Homem de Mello de Souza HA, Dalla-Costa LM, Vicenzi FJ, Camargo de Souza D, Riedi CA, Filho NA, et al. MALDI-TOF: a useful tool for laboratory identification of uncommon glucose non-fermenting Gram-negative bacteria associated with cystic fibrosis. J Med Microbiol 2014;63:1148−53.

[31] Leuschner RG, Beresford-Jones N, Robinson C. Difference and consensus of whole cell *Salmonella enterica* subsp. enterica serovars matrix-assisted laser desorption/ionization time-of-flight mass spectrometry spectra. Lett Appl Microbiol 2004;38:24−31.

[32] Schulthess B, Brodner K, Bloemberg GV, Zbinden R, Böttger EC, Hombach M. Identification of Gram negative cocci by use of MALDI-TOF mass spectrometry: comparison of different preparation methods and implementation of a practical algorithm for routine diagnostics. J Clin Microbiol 2013;51:1834−40.

[33] TeKippe EM, Shuey S, Winkler DW, Butler M, Burnham CAD. Optimizing identification of clinically relevant Gram-positive organisms by use of the Bruker Biotyper matrix-assisted laser desorption ionization-time of flight mass spectrometry system. J Clin Microbiol 2013;51:1421−7.

[34] Pascual Hernández A, Ballestero-Téllez M, Galán-Sánchez F, Rodríguez Iglesias M. Aplicación de la espectrometría de masas en la identificación de bacterias. Enferm Infecc Microbiol Clin 2016;34(Suppl 2):8−18.

[35] Dupont C, Sivadon-Tardy V, Bille E, Dauphin B, Beretti JL, Alvarez AS, et al. Identification of clinical coagulase-negative staphylococci, isolated in microbiology laboratories, by matrix-assisted laser desorption/ionization-time of flight mass spectrometry and two automated systems. Clin Microbiol Infect 2010;16:998−1004.

[36] Kassim A, Pflüger V, Premji Z, Daubenberger C, Revathi G. Comparison of biomarker based matrix assisted laser desorption ionization-time of flight mass spectrometry (MALDI-TOF MS) and conventional methods in the identification of clinically relevant bacteria and yeast. BMC Microbiol 2017;17:128.

[37] Zhou C, Hu B, Zhang X, Huang S, Shan Y, Ye X. The value of matrix-assisted laser desorption/ionization time-of-flight mass spectrometry in identifying clinically relevant bacteria: a comparison with automated microbiology system. J Thorac Dis 2014;6:545−5552.

[38] Chen JHK, She KK, Wong O-Y, Teng JLL, Yam W-C, Lau SKP, et al. Use of MALDI Biotyper plus ClinProTools mass spectra analysis for correct identification of *Streptococcus pneumoniae* and *Streptococcus mitis/oralis*. J Clin Pathol 2015;68:652−6.

[39] Richter SS, Sercia L, Branda JA, Burnham CA, Bythrow M, Ferraro MJ, et al. Identification of *Enterobacteriaceae* by matrix-assisted laser desorption/ionization time-of-flight mass spectrometry using the VITEK MS system. Eur J Clin Microbiol Infect Dis 2013;32:1571−8.

[40] Abrok M, Arcson A, Lazar A, Urban E, Deak J. Combination of selective enrichment and MALDI-TOF MS for rapid detection of *Streptococcus agalactiae* colonisation of pregnant women. J Microbiol Methods 2015;114:23−5.

[41] Fang H, Ohlsson A-K, Ullberg M, Ozenci V. Evaluation of species-specific PCR, Bruker MS, VITEK MS and the VITEK 2 system for the identification of clinical *Enterococcus* isolates. Eur J Clin Microbiol Infect Dis 2012;31:3073—7.

[42] Vila J, Juiz P, Salas C, Almela M, de la Fuente CG, Zboromyrska Y, et al. Identification of clinically relevant *Corynebacterium* spp., *Arcanobacterium haemolyticum*, and *Rhodococcus equi* by matrix-assisted laser desorption ionization-time of flight mass spectrometry. J Clin Microbiol 2012;50:1745—7.

[43] Siller Ruiz M, Hernández Egido S, Calvo Sánchez N, Muñoz Bellido JL. Unusual clinical presentations of *Actinotignum (Actinobaculum) schaalii* infection. Enferm Infecc Microbiol Clin 2017;35:197—8.

[44] Quiles Melero I, Pelaez T, Rezusta Lopez A, Garcia-Rodriguez J. Aplicacion de la espectrometria de masas en micologia. Enferm Infecc Microbiol Clin 2016;34(Suppl 2):26—30.

[45] Pence MA, McElvania TeKippe E, Wallace MA, Burnham CAD. Comparison and optimization of two MALDI-TOF MS platforms for the identification of medically relevant yeast species. Eur J Clin Microbiol Infect Dis 2014;33:1703—12.

[46] Zboromyrska Y, Rubio E, Alejo I, Vergara A, Mons A, Campo I, et al. Development of a new protocol for rapid bacterial identification and susceptibility testing directly from urine samples. Clin Microbiol Infect 2016;22(561):e1—6.

[47] Wang XH, Zhang G, Fan YY, Yang X, Sui WJ, Lu XX. Direct identification of bacteria causing urinary tract infections by combining matrix-assisted laser desorption ionization-time of flight mass spectrometry with UF-1000i urine flow cytometry. J Microbiol Methods 2013;92:231—5.

[48] Sánchez-Juanes F, Siller Ruiz M, Moreno Obregón F, Criado González M, Hernández Egido S, de Frutos Serna M, et al. Pretreatment of urine samples with SDS improves direct identification of urinary tract pathogens with matrix-assisted laser desorption ionization-time of flight mass spectrometry. J Clin Microbiol 2014;52:335—8.

[49] March Rosselló GA, Gutiérrez Rodríguez MP, Ortiz de Lejarazu Leonardo R, Orduña Domingo A, Bratos Pérez MA. New procedure for rapid identification of microorganisms causing urinary tract infection from urine samples by mass spectrometry (MALDI-TOF). Enferm Infecc Microbiol Clin 2015;33:89—94.

[50] Íñigo M, Coello, A, Fernández-Rivas G, Rivaya B, Hidalgo J, M.D. Quesada, et al. Direct identification of urinary tract pathogens from urine samples, combining urine screening methods and matrix-assisted laser desorption ionization-time of flight mass spectrometry, J Clin Microbiol 2016;54:988—93.

[51] Oviaño M, Ramírez CL, Barbeyto LP, Bou G. Rapid direct detection of carbapenemase-producing Enterobacteriaceae in clinical urine samples by MALDI-TOF MS analysis. J Antimicrob Chemother 2017;72:1350—4.

[52] Lupo A, Papp-Wallace KM, Sendi P, Bonomo RA, Endimiani A. Non-phenotypic tests to detect and characterize antibiotic resistance mechanisms in *Enterobacteriaceae*. Diagn Microbiol Infect Dis 2013;77:179—94.

[53] Sparbier K, Lange C, Jung J, Wieser A, Schubert S, Kostrzewa M. MALDI biotyper-based rapid resistance detection by stable-isotope labeling. J Clin Microbiol 2013;51:3741—8.

[54] Lange C, Schubert S, Jung J, Kostrzewa M, Sparbier K. Quantitative MALDI-TOF MS for rapid resistance detection. J Clin Microbiol 2014;52:4155—62.

[55] Jung JS, Hamacher C, Gross B, Sparbier K, Lange C, Kostrzewa M, et al. Evaluation of a semiquantitative matrix-assisted laser desorption ionization-time of flight mass spectrometry method for rapid antimicrobial susceptibility testing of positive blood cultures. J Clin Microbiol 2016;54:2820—4.

CHAPTER ELEVEN

Direct Application of MALDI-TOF Mass Spectrometry to Cerebrospinal Fluid for Pathogen Identification

Ana Lara Oya
Department of Microbiology, Complejo Hospitalario de Jaén, Jaén, Spain

11.1 INTRODUCTION

11.1.1 Acute bacterial meningitis

Acute bacterial meningitis is the most common infection of the central nervous system and represents one of the most serious and clinically significant manifestations of bacterial infection. It is a medical emergency requiring a rapid diagnosis and an immediate treatment. Delay in appropriate antibiotic treatment increases mortality and neurological sequelae [1,2].

The infections can either be community-acquired or hospital-acquired, e.g., after neurosurgical intervention, as a complication of severe neurotrauma or related to indwelling cerebrospinal fluid (CSF) drains. Community-acquired bacterial meningitis is associated with high mortality and morbidity rates. The distribution of pathogens varies according to the patient's age and status of the immune system.

Neonatal meningitis is typically caused by *Streptococcus agalactiae* (Group B streptococcus) and *Escherichia coli*. For children after the neonatal period and adults, the majority of meningitis cases are caused by *Streptococcus pneumoniae* (pneumococcus), *Neisseria meningitidis* (meningococcus) [3,4], and *Haemophilus influenzae*. The incidence of bacterial meningitis caused by *H. influenzae* in this group has decreased in recent years due to the use of *H. influenzae* Type B vaccine. *Listeria monocytogenes* is the third most common cause of bacterial meningitis in general, but

The Use of Mass Spectrometry Technology (MALDI-TOF) in Clinical Microbiology.
DOI: https://doi.org/10.1016/B978-0-12-814451-0.00011-3

in specific risk groups the proportion of cases due to listeria is considerably higher, such as the elderly and immunocompromised patients, patients with cancer, and patients using immunosuppressive therapy or diabetics [5].

Diagnosis of acute bacterial meningitis is classically based on the combination of typical clinical symptoms and the laboratory tests. Conventionally, pathogens responsible for bacterial meningitis have been detected by Gram staining of CSF.

Samples generating negative smears are cultured to rule out the presence of circulating bacteria, while patients with positive cultures are treated immediately with broad-spectrum antibiotics based upon the Gram stain result until a definitive identification based on culture. Molecular methods are powerful but they are still expensive and require sophisticated technology and expertise personnel [6,7].

11.1.2 Application of matrix-assisted laser desorption/ ionization—time-of-flight mass spectrometry in clinical samples

Recently, the development of matrix-assisted laser desorption/ionization—time-of-flight mass spectrometry (MALDI-TOF MS) has revolutionized the identification of bacteria, yeast, and molds in clinical microbiology by introducing simple, reliable, rapid, and low-cost techniques.

This technology might be directly applied to some clinical samples, such as blood, urine, CSF, pleural fluid, peritoneal liquid, and synovial fluid. Most commonly, MALDI-TOF MS is used to identify microorganisms that grow from positive blood cultures. Due to the high microorganism burden in these bottles, they may be acceptable specimens for MALDI-TOF MS identification. The only other specimen that has been attempted with any success thus far is the urine.

The presence or absence of specific proteins in patient's CSF was used as a biomarker for a number of neurological disorders [8,9], but current guidelines do not address MALDI-TOF MS in the diagnosis of bacterial meningitis [3,10]. There have been many unsuccessful attempts for identifying pathogens directly on the CSF by various authors. Some of them claim that direct MALDI-TOF MS testing on CSF samples appears to be of no benefit in accelerating the microbiological diagnosis of community-acquired meningitis and healthcare-associated meningitis [7].

Nevertheless, there are very few reports describing the successful use of MALDI-TOF MS for the direct detection of microorganisms causing bacterial meningitis [7,11,12].

11.2 PERFORMANCE IN ROUTINE CLINICAL MICROBIOLOGY

11.2.1 CSF preparation

Following sample acquisition, in general the CSF was subjected to low-speed centrifugation to remove leukocytes and then followed by high-speed centrifugation to pellet bacteria. The pelleted debris was then solubilized in formic acid–acetonitrile, centrifuged, and analyzed by MALDI-TOF MS using the BioTyper database. Only three protocols have been specifically described in previous medical publications [7,11,12]. These three protocols are as follows:

1. Centrifuge 1000 μL CSF for 5 minutes at 13,500 × g. Blood cell components, mostly leukocytes, are removed by aspiration, and the resulting pellets is used for MALDI-TOF MS analysis. Wash the pellets with 300 μL of high-pressure liquid chromatography (HPLC) deionized water and 900 μL of absolute ethanol and centrifuge again for 5 minutes at 13,500 × g. Treat the dried pellets with 50 μL of 70% formic acid and 50 μL of 100% acetonitrile for 1 minute to extract bacterial cell contents. Centrifuge the extract for 5 minutes at 13,500 × g. The aliquots of the supernatant are subjected to bacterial identification by MALDI-TOF MS.

2. Centrifuge CSF for 30 seconds at 2000 rpm to remove leukocytes. Then, centrifuge the supernatant at 13,000 rpm for 5 minutes to collect bacteria. Dry the pellet and wash once with 0.5 mL HPLC deionized water. Then, centrifuge the supernatant again at 13,000 rpm for 5 minutes to form a pellet. The cell contents are extracted from the pellet using equal amounts of 70% formic acid and acetonitrile. Centrifuge the extract again at 13,000 rpm for 2 minutes and apply 1 μL to a MALDI-TOF steel target plate and left to dry.

3. Centrifuge 1 mL CSF for 30 seconds at 2000 rpm. Centrifuge the supernatant at 13,000 rpm for 5 minutes and wash the resulting pellet with 1 mL of HPLC deionized water. Then, centrifuge the suspension

at 13,000 rpm for 5 minutes and remove the supernatant. Treat the dried pellet with equal amounts of 70% formic acid and acetonitrile. Centrifuge the extract again at 13,000 rpm for 2 minutes and apply 1 mL of aliquot from the supernatant to the MALDI-TOF MS target plate and left to dry.

11.2.2 MALDI-TOF MS analysis

The MALDI-TOS MS analysis should be elaborated as recommended by the manufacturer:

Overlap the spots with 1 µL matrix solution (alpha-cyano-4-hydroxy-cinnamic acid; HCCA), which was a saturated solution of HCCA in 50% acetonitrile and 2.5% trifluoroacetic acid. After drying at room temperature to enable co-crystallization, analyze the target plate by the MALDI-TOF MS. The peak lists generated were used to find matches against the reference library by direct use of the integrated pattern-matching algorithm of the software. The score, number of identical identifications, distance to the next best taxon match, and the automatic evaluation of the obtained identification have to be taken into consideration to validate the final result.

11.3 LIMITATIONS

In general, for microorganisms MALDI-TOF MS identification in direct clinical samples must consider the following limitations [13]:
- Bacterial load: It is difficult to use this approach in samples with low microbial density.
- Limited volume of sample available.
- Substances that may interfere in the analysis of spectra (e.g., cell debris, different proteins in biological fluids, meconium remains in amniotic fluid, etc.) and that are not completely eliminated during the repair process the sample.

Finally, polymicrobial samples (although CSF samples are not common) are not suitable for direct identification, because there are no tools available to discriminate more than one microorganism in the same spectrum and purulent samples are also problematic.

Bacterial identification from CSF is strongly limited by the low bacterial load and the limited volume available. For these reasons, it is yet not applicable in routine diagnostic laboratories.

However, samples obtained through external ventricular drainages allow direct identification with MALDI-TOF-MS with high probability of success in case of infection. This is explained by two main reasons: there is a higher sample volume and bacterial density is usually high, due mainly to the colonization of the internal surface of the drainage tube.

11.4 FUTURE DIRECTIONS

MALDI-TOF MS is a promising tool for the rapid identification of pathogens in patients with monomicrobial bacterial meningitis. Currently, there are few publications describing the identification direct of microorganisms in CSF. *Streptococcus pneumoniae* has been identified in the CSF in a patient with bacterial meningitis [11] and also *Klebsiella pneumoniae* in a patient who developed bacterial meningitis 2 weeks later of being submitted to a surgical intervention of a brain tumor [7]. In 2011, at the annual congress of the ESCMID (European Society of Clinical Microbiology and Infectious Diseases) a paper was presented in which the authors obtained the spectra of 183 consecutive samples of CSF by MALDI-TOF MS; 14 of them were culture positive [14]. In this case, no valid identification was obtained. Recently, Bishop et al. [12] have published a study which assesses MALDI-TOF MS sensitivity in patients with clinical meningitis and a positive CSF Gram stain. In this case, MALDI-TOF MS might be helpful in directing appropriate antibiotic treatment while awaiting culture results in centers where Gram-negative bacteria are prevalent. Nevertheless, according to their results, direct MALDI-TOF MS testing on CSF samples appears to be of no benefit in accelerating the microbiological diagnosis of community-acquired meningitis and healthcare-associated staphylococcal meningitis, and in these cases Gram stains might be sufficient to direct empirical antibiotic treatment.

11.5 CONCLUSION

The use of this technique in combination with Gram stain and traditional bacterial culture could represent an important turning point in the diagnosis of bacterial meningitis, increasing sensitivity, and decreasing time to diagnosis and allowing for targeted and aggressive antibiotic therapy for a patient population that is critically ill. In order to be able to identify microorganisms directly from biological fluids such as CSF, a high density of microorganisms is needed and, in addition, the limited volume of CSF that usually reaches at clinical laboratory is an important limitation. Currently, new extraction and concentration sample protocols would be needed to increase their yield and to be able to obtain a correct amount of proteins for the generation of a quality mass spectrum. This is an application of MALDI-TOF MS that warrants significant further investigation.

REFERENCES

[1] Lepur D, Baršić B. Community-acquired bacterial meningitis in adults: antibiotic timing in disease course and outcome. Infection 2007;35:225−31. Available from: https://doi.org/10.1007/s15010-007-6202-0.

[2] Aronin SI. Community-acquired bacterial meningitis: risk stratification for adverse clinical outcome and effect of antibiotic timing. Ann Intern Med 1998;129:862. Available from: https://doi.org/10.7326/0003-4819-129-11_Part_1-199812010-00004.

[3] van de Beek D, Cabellos C, Dzupova O, Esposito S, Klein M, Kloek AT, et al. ESCMID guideline: diagnosis and treatment of acute bacterial meningitis. Clin Microbiol Infect 2016;22(Suppl 3):S37−62. Available from: https://doi.org/10.1016/j.cmi.2016.01.007.

[4] Brouwer MC, van de Beek D. Management of bacterial central nervous system infections. Handb Clin Neurol 2017;140:349−64. Available from: https://doi.org/10.1016/B978-0-444-63600-3.00019-2.

[5] van Ettekoven CN, van de Beek D, Brouwer MC. Update on community-acquired bacterial meningitis: guidance and challenges. Clin Microbiol Infect 2017;23:601−6. Available from: https://doi.org/10.1016/j.cmi.2017.04.019.

[6] Kumar A, Debata PK, Ranjan A, Gaind R. The role and reliability of rapid bedside diagnostic test in early diagnosis and treatment of bacterial meningitis. Indian J Pediatr 2015;82:311−14. Available from: https://doi.org/10.1007/s12098-014-1357-z.

[7] Segawa S, Sawai S, Murata S, Nishimura M, Beppu M, Sogawa K, et al. Direct application of MALDI-TOF mass spectrometry to cerebrospinal fluid for rapid pathogen identification in a patient with bacterial meningitis. Clin Chim Acta 2014;435:59−61. Available from: https://doi.org/10.1016/j.cca.2014.04.024.

[8] Taibi L, Boursier C, Clodic G, Bolbach G, Bénéteau-Burnat B, Vaubourdolle M, et al. Search for biomarkers of neurosarcoidosis by proteomic analysis of cerebrospinal fluid. Ann Biol Clin (Paris) 2017;75:393−402. Available from: https://doi.org/10.1684/abc.2017.1260.

[9] Fania C, Arosio B, Capitanio D, Torretta E, Gussago C, Ferri E, et al. Protein signature in cerebrospinal fluid and serum of Alzheimer's disease patients: the case of apolipoprotein A-1 proteoforms. PLoS One 2017;12. Available from: https://doi.org/10.1371/journal.pone.0179280 e0179280.

[10] Tunkel AR, Hartman BJ, Kaplan SL, Kaufman BA, Roos KL, Scheld WM, et al. Practice guidelines for the management of bacterial meningitis. Clin Infect Dis 2004;39:1267−84. Available from: https://doi.org/10.1086/425368.

[11] Nyvang Hartmeyer G, Kvistholm Jensen A, Böcher S, Damkjaer Bartels M, Pedersen M, Engell Clausen M, et al. Mass spectrometry: pneumococcal meningitis verified and Brucella species identified in less than half an hour. Scand J Infect Dis 2010;42:716−18. Available from: https://doi.org/10.3109/00365541003754493.

[12] Bishop B, Geffen Y, Plaut A, Kassis O, Bitterman R, Paul M, et al. The use of matrix-assisted laser desorption ionization time-of-flight mass spectrometry (MALDI-TOF-MS) for rapid bacterial identification in patients with smear-positive bacterial meningitis. Clin Microbiol Infect 2017. Available from: https://doi.org/10.1016/j.cmi.2017.05.014.

[13] Vila J, Zboromyrska Y, Burillo A, Bouza E. Perspectivas de futuro de la espectrometría de masas en microbiología. Enferm Infecc Microbiol Clin 2016;34:53−8. Available from: https://doi.org/10.1016/S0213-005X(16)30192-6.

[14] Bjørnholt V, Nilsen SM, Noorland I, Wigemyr M, Løken CH. MALDI-TOF mass spectrometry ID of bacteria directly from cerebrospinal fluid. What you see is what you get. Clin Microbiol Infect 2011;17:S66 21st Eur Congr Clin Microbiol Infect Dis:O-345.

CHAPTER TWELVE

Application of MALDI-TOF Mass Spectrometry in Clinical Virology

Cristina Gómez Camarasa and Fernando Cobo

Department of Microbiology, University Hospital Virgen de las Nieves, Granada, Spain

12.1 INTRODUCTION

The matrix-assisted laser desorption ionization−time-of-flight mass spectrometry (MALDI-TOF MS) is an analytical method of microbial identification and characterization based on the fast and precise assessment of the mass of molecules in a variable range of 100 Da−100 KDa.

The first description of the use of MS technology for bacterial identification was in 1975 [1]. Furthermore, MALDI-TOF MS was used in the mid-1990s for the identification of bacteria in research settings [2,3] but it took a long time for the introduction of this technology in routine microbiology. Then, in 2004, the first complete database for bacterial identification was reported [4].

The possibility of MALDI-TOF to quickly characterize a wide variety of microorganisms including bacteria, fungi, and viruses [5] increases its potential application in some areas of microbiological diagnosis. Thus, this technology can be used for rapid microbial identification at a relatively low cost, and it is an alternative for conventional laboratory diagnosis and molecular identification systems. The major contribution of this method is that it has drastically improved the time to identification of a positive sample.

The main use of this technology has been, until now, the identification or typing of bacteria from a positive culture. However, applications in the field of clinical virology have been also performed and are opening up new opportunities although the introduction in the routine laboratory diagnosis still remains a challenge.

The Use of Mass Spectrometry Technology (MALDI-TOF) in Clinical Microbiology.
DOI: https://doi.org/10.1016/B978-0-12-814451-0.00012-5

In this review, MALDI-TOF technique is presented including sample preparation as well as data management and quality control, and mainly focused in viral research and diagnosis.

12.2 MS PRINCIPLES

The mass spectrometer is composed of three main components: an ion source to ionize and transfer sample molecules ions into a gas phase, a mass analyzer device that separate molecules depending to their mass, and finally a detector to monitor all separated ions. There are several ionization methods such as chemical ionization, electrospray, and MALDI. The choice of method basically depends to the objective of the MS analysis and the nature of the sample, although both MALDI and electrospray are techniques that permit ionization and vaporization of a big quantity of biomolecules [6].

In MS technique [7], samples are prepared by mixing several biomolecules and embedded in a matrix that will crystallize inside. The matrix is composed of small acid molecules that have a strong optical absorption in the range of the laser wavelength used, although the composition is different according to the type of laser used and the biomolecule to be analyzed. The molecules are then both desorbed and ionized by charge transfer by absorbing the energy of a short laser pulse. The desorbed and ionized molecules are first accelerated in an electrical field and ejected through a metal flight tube and collide with a detector at the end of the flight tube. This cycle is repeated at a frequency of 50 Hz or more. Thus, biomolecules are separated according to their mass and create a mass spectrum that is characterized by both the mass and the intensity of the ions. Finally, the result is a spectral signature (spike representation) that is then searched for in the appropriate database for the identification of the microorganism, comparing with values provides by the manufacturer of the database.

On the other hand, electrospray ionization mass spectrometry (ESI-MS) generally uses DNA amplicons that are dissolved in a solvent and then injected in a conductive capillary. Inside this capillary, high voltage is applied resulting in the emission of aerosols of charged drops of the sample. After this, the sample is sprayed through compartments with diminishing pressure, resulting in the formation of gas-phase multiple-charged analyte ions, which are finally detected by the spectrometer. ESI–MS technology permits high resolution analyzing the basic composition of amplicons [8].

There are also two available methods for typing using this technology. The first method is the MALDI-resequencing that uses the iSeq method [9,10]. In this technique, RNA strands are cleaved and the products of cleavage are analyzed by MS, and their spectra are compared with the spectra of the reference sequences. Initially, MALDI-RE was used mainly for the detection of single-nucleotide polymorphisms, but currently there are additional applications for this method [11]. The second method for typing is based on the use of ESI technology to analyze specific DNA amplicons (PCR–ESI–MS) [12].

12.3 CURRENT CLINICAL VIROLOGY APPLICATIONS

12.3.1 Viral diagnosis

Viruses were traditionally detected by cell culture, which in spite of being gold standard, often took days or even weeks before any results were available. Common laboratory techniques in viruses' detection include antibody analysis such as enzyme-linked immunosorbent assay, molecular techniques like polymerase chain reaction (PCR) and dot blot hybridization. However, MALDI-TOF technology has been already used to investigate a wide variety of viruses clinically relevant but it has advanced less as it has in bacteriology or mycology. There are several approaches of this technology for application in the field of virology.

A recent study performed a multiplex MALDI-TOF MS method that can successfully detect human herpes viruses from a wide variety of archival biological specimens [13]. The concordance rate between the MALDI-TOF MS technique and reference methods was high and the detection limits of the MS methods were comparable to the previous reports of multiplex herpes virus detection using an oligonucleotide microarray and multiplex PCR techniques. Low viral loads near the detection limit of the method and weak cross-reactivity might contribute to the few discrepant results with reference methods. Thus, this multiplex approach can be very useful for large-scale epidemiological research studies in which broad viruses detection is necessary and may also become highly useful for multiplex clinical diagnostic testing.

The rapid identification of existing and emerging respiratory viruses is crucial in combating outbreaks and epidemics. The detection of respiratory viruses such as influenza viruses are mainly performed using PCR techniques (real-time reverse transcription polymerase chain reaction; rRT-PCR) or antibody-based assays to identify the nucleoproteins. Alternative methods have been investigated for viral diagnosis, such as multiplexed flow cytometry [14], microarrays [15], and MS [16]. MS has been considered as one of the best methods for protein analysis because of its low detection limit and high accuracy. It has been also demonstrated that the combined use of antibody-magnetic nanoparticles and MALDI-TOF MS is suitable for sensitive detection of influenza viruses [17]. This method could also be used for the rapid screening of virus subtypes, suggesting a promising application in early and accurate diagnosis of influenza viruses. It has developed rapid sample preparation method to allow identification of cultured respiratory viruses with MS and multiplex PCR coupled with this technique to detect the most common respiratory viruses such as influenza virus, parainfluenza viruses, coronavirus, metapneumovirus, respiratory syncytial virus, adenovirus, and bocavirus [18–20].

Enteroviruses are also a major cause of illness in children causing gastroenteritis, hand, foot, and mouth disease and severe neurological complications such as aseptic meningitis, encephalitis, and acute neurogenic pulmonary edema. Classical detection of enteric virus is mostly based on conventional approaches such as indirect immunofluorescence assay, neutralizing testing, and cell culture and immunohistochemical detection. However, some clinical samples remain undetectable because of the absence of antibodies against viral proteins, and cell culture and neutralization tests are time-consuming and laborious. Multiplex RT-PCR and real-time PCR have shown good sensitivity and specificity [21]. For these viruses, the MALDI-TOF MS technology has demonstrated high capacity to diagnose with high throughput and short turnaround time. MALDI-TOF MS measures the intrinsic physical properties of molecules directly. A PCR mass assay combining multiplex PCR with MALDI-TOF MS technology has shown simultaneous detection of eight human enteric viruses including poliovirus, coxsackievirus A16, enterovirus 71, hepatitis E virus, echovirus, norovirus, astrovirus, and reovirus [22]. This method has a sensitivity ranging from 100 to 1000 copies per reaction and shows that this assay could be used for the simultaneous detection of co-infections with several viruses.

MALDI-TOF MS has been also used for the diagnosis of hand, foot, and mouth disease that is caused by acute enterovirus infections such as poliovirus, coxsackievirus A and B, and echovirus. These infections are clinically indistinguishable but infection with echovirus 71 could be complicated with aseptic meningitis, encephalitis, and acute flaccid paralysis. Recently, a study has developed a MALDI-TOF MS technique for the rapid diagnosis of these viruses [23]. The diagnosis of these pathogens can help to a more comprehensive surveillance of enteroviruses to understand the molecular epidemiology of enteroviruses detecting multiple genotypes simultaneously. The main advantage of this strategy is the identification of several genotypes at the same time, opening up new possibilities for the diagnosis and discovering pathological and epidemiological characteristics.

On the other hand, infection with high-risk human papillomaviruses (HPV) has been closely associated with cervical cancer as well as other types of cancer such as anus, vagina, penis, and head and neck. The diagnosis of HPV infection is based on several methods, but HPV DNA detection is necessary for the follow-up of the patients. Currently, there are three HPV assays approved for clinical use by the US Food and Drug Administration: two non-PCR-based assays (e.g., Cervista and Hologic, Marlborough, MA; Hybrid Capture 2, Qiagen, Valencia, CA) and a PCR-based method such as the Cobas HPV assay. The majority of these assays use GP5 + /GP6 + consensus primers, MY09/11 primers, and their multiprimer derivatives PGMY09/11 primers and the SPF10 system. However, a new method based on the amplification of a fragment of the L1 gene region using GP5 + /GP6 + consensus primers using mass array technique based on the MALDI-TOF MS technology has been recently developed to identify 14 high-risk HPVs [24–26]. This MS assay was found to be non-cross-reactive with common low-risk HPV genotypes affecting the genital tract and more sensitive (General sensitivity results ranging from 10 to 100 copies) than the Hybrid Capture 2 and conventional PCR methods. An advantage of this method is that it is possible to extend the HPV genotype spectrum by adding new type-specific primers. The use of this technique together with automated DNA extraction and PCR pipetting procedures could increase the viral detection which a decrease in the cost of the overall assay.

12.3.1.1 Diagnosis of mutations and identification of viral genotypes

Hepatitis B virus (HBV) and hepatitis C virus (HCV) infections are still a major public health problem due to their worldwide morbidity and mortality. For this reason, screening for these infections is an important part of routine laboratory activity. Serological and molecular markers are key elements for the diagnosis, prognosis, and monitoring of treatment in HBV and HCV infections. Currently, chemiluminescence assays are widely being used for viral diagnosis. Molecular techniques are routinely used for detecting, quantifying viral genomes, and analyzing their sequence to determine their genotype and detect resistance to antiviral drugs. However, MassARRAY systems based on nucleic acid analysis by MALDI-TOF MS technology provide an alternative approach to HBV and HCV genotyping [27], being accurate to identify all eight HBV genotypes. MassARRAY system generates MS patterns that are compared to sequences derived from known HBV genotypes. The main advantages of this assay are its rapidity, cost-effectiveness, and versatility. The assay produces automatic data reports and does not require any additional data interpretation.

Moreover, this method can discriminate single-point mutations and is capable of detecting wild-type and mutant alleles [28,29]. Variations in the HBV genome are biologically and clinically important and might confer drug resistance or affect virus replication capacity, resulting in failure of antiviral therapy. Some MALDI-TOF MS platforms are capable to detect 60 HBV variants with accuracy and low detection limits [29]. This technology has been used also to study the evolution of drug resistance and to detect minor HBV variants [30]. The main disadvantage of this method is that only HBV genotypes B and C were tested.

Some characteristic mutations in HBV could cause lamivudine resistance due to prolonged treatment with this antiviral agent. Current methods for detecting such variants (e.g., sequencing analysis, restriction fragment length polymorphism (RFLP), and hybridization-based assays) are time-consuming and unsuitable for screening large numbers of samples. For these reasons, it has been developed a MALDI-TOF MS assay suitable for detecting YMDD mutants [28]. This technique provides high rates of accuracy for YMDD mutations and has advantages over RFLP and direct DNA sequencing, including the ability to more sensitively and specifically detects mixed populations as well as being a simpler procedure without need for cloning or multiple PCR steps producing earlier detection of HBV mutations. Moreover, this platform is adaptable for the detection of other mutations or polymorphisms. This method will be very useful to evaluate the emergence of mutant viruses and to monitoring of antiviral treatment of chronic hepatitis B, correlating them to clinical features with regard to response to drug therapy in these patients.

On the other hand, the identification of HCV genotypes has become important to determine the clinical course and the outcome of antiviral therapy. Many HCV genotyping methods have been developed, including genotype-specific PCR, RFLP, hybridization techniques, and heteroduplex motility assays. Currently, the method of choice for this purpose is the nucleotide sequencing of a subgenomic region, but all these methods are complex to standardize, labor-intensive, or non-useful for determining multiple HCV genotypes. However, it is frequent heterogeneity during the course of infection, producing mixed-genotypes infections in some patients. Thus, researchers have developed some MALDI-TOF MS–based assays for HCV genotyping during the last years [31–33]. For genotyping purposes, MS provides accurate information on the molecular mass of the analyte and the procedure can be fully automated, as well as the direct analysis of molecules with no labels to be introduced and no separation steps followed by the image processing.

Other methods based on MALDI-TOF MS technology for viral diagnosis purposes have been also recently developed. A method based on MALDI-TOF MS technology studied the possibility of genotyping JC virus using urine samples in patients without clinical evidence of progressive multifocal leukoencephalopathy [34]. MALDI-TOF showed successfully genotype of these viruses and identifies sequence variations in the target genome. Another study used this technique to identify the possible mutations of influenza A (H5) viruses for monitoring purposes [35].

12.3.1.2 Identification of drug resistance
Finally, the application of MALDI-TOF MS technique could be useful for detecting drug resistance against some antivirals. A recent study has developed a sensitive and rapid method for the detection of ganciclovir resistance by PCR-based MALDI-TOF analysis [36].

12.3.1.3 Use of MALDI-TOF techniques in epidemiology
Other potential application of MALDI-TOF MS technology is to provide critical epidemiological data about viral infections. From an epidemiological point of view, one of the main applications of these methods is determining specific data of clinical isolates for viral diseases' outbreaks. Because of their complexity, some viral typing techniques require specialized test methods and additional diagnostic devices. For this reason, the typing is not routinely performed in every laboratory, so these samples are being sent out of the hospital and this fact could delay the investigation of

Table 12.1 Summary of the main applications of MALDI-TOF MS to clinical virology
Applications

Identification of viruses from clinical samples
Diagnosis of mutations for viruses
Screening of viral subtypes
Identification of antiviral resistance
Epidemiology of viral infections

these outbreaks and hospital-associated infections. The introduction of MALDI-TOF MS technology in the laboratories can facilitate the performing of tests leading to improve the information to the infection control staff. The introduction of MS techniques into the clinical laboratory can provide accurate and rapid epidemiological data, and it could have a critical role in hospital infection control and surveillance of viral outbreaks (Table 12.1).

12.4 ADVANTAGES OF MALDI-TOF MS TECHNOLOGY AND COST-EFFECTIVENESS

With regard to viral diagnosis, there are few studies comparing MALDI-TOF MS technology with other conventional based methods (e.g., viral, culture, nucleic acid–based techniques). MALDI-TOF method successfully detects viruses in a wide variety of biological specimens, and the concordance rate between MALDI-TOF and these techniques is high. PCR-based identification strategies have some limitations such as much longer turnaround time than MS, reagent and labor costs, and some technical problems. On the other hand, MALDI-TOF technology can also be used to identify viruses in co-infected samples. These methods are capable of simultaneous detection of multiple pathogens in a single assay and can then prevent misdiagnosis and delays in treatment without increasing costs or adding a new step to the process.

With regard to the technical problems, PCR methods could have inhibitory particles and issues of contamination; thus, separate areas for sample preparation, amplification, and analysis are needed. Furthermore, nucleic acid–based techniques are expensive and time-consuming and appear in most cases less convenient than MS for routine laboratory identifications. For MALDI-TOF there must be significant progress made in closing the upfront of PCR processing to include DNA extraction and the PCR process itself.

MALDI-TOF MS methods also allow large-scale research studies not only for fresh samples but also on archival samples from several biological specimens.

The main advantages and disadvantages of MALDI-TOF MS technology are shown in Table 12.2.

The most remarkable differences between MALDI-TOF method and other conventional techniques are the estimated time and costs required for sample identification. There are no studies of cost-effectiveness in viruses, but it is estimated that the cost of bacterial identification represents only 17%−32% of the costs of conventional identification methods [37]. Other studies have shown that the cost of reagents required for phenotypic and nucleic acid identification is higher than for MS techniques [38]. The expensive cost of MALDI-TOF devices is comparable with other laboratory equipment. Moreover, MALDI-TOF is a rapid technique if compared with other methods, so this technology improves the working time (e.g., preanalytical procedure) as well as the turnaround time (e.g., analytical procedure). The time needed for microorganism identification is <15 minutes for MS method, whereas conventional techniques need 5−48 hours.

One of the main weaknesses of MALDI-TOF is that the identification is limited by database. The mass profile is used as a mass spectrum to compare with those of well-characterized organisms in a database. The spectrum typically includes specific peaks, so that with a comprehensive library of spectra, the identification might be performed by using bioinformatics.

Table 12.2 Advantages and disadvantages of MALDI-TOF MS technology for viral analysis compared to other methods

Advantages	Disadvantages
Characterization of a wide variety of viruses	Identification limited by database for sequencing of genetic markers:
Reliable and rapid identification	
Accurate and sensitive method	
High throughput	• Time-consuming
Easy sample preparation	• High cost
Cost-effective	• Different genetic markers
Improve the patient management	
Automation possible	

Currently, there are a wide number of references in the database from the available commercial platforms for MALDI-TOF techniques. These mass spectra can establish the comparison and further identification of microorganisms, but until now these spectra are limited. The database will expand due to the collaboration between commercial companies and the hospitals of several countries, so the available databases should be improved in the near future. The comparison of data from different companies' instruments will be very important, but this fact is currently not feasible.

12.5 DATA MANAGEMENT AND QUALITY CONTROL

Sample preparation by MALDI-TOF MS remains laborious today, although some robots are being developed to improve the deposit of the sample and the matrix on the target. Moreover, some aspects of MALDI-TOF MS technology are being also investigated; for example, the laser pulse frequency, disposable target slides, and the mass spectrometers as well as the software used for the communication between laboratory equipment and laboratory information systems (LIS).

Data management primarily depends on the interpretation of crude data. Currently, some methods for data interpretation have been investigated [39]. An important fact is the integration of information communication systems to physicians that requires hospital information systems or LIS.

Reference databases require an update if new information is obtained. The addition of new isolates or mutations could improve this tool and the final diagnosis of the diseases. A robust system of information for regional prevalence of infectious diseases, emerging and reemerging infectious pathogens, the discovery of new mutations, and local or pandemic changes in epidemiology is necessary.

The MALDI-TOF MS system is increasingly used in clinical diagnostic laboratory for viral identification, mutation analysis, genotyping, and antiviral resistance studies. However, there are still some problems arising during sample extraction affecting to the extraction protocol and to the adequate reagent conservation for the assay. These problems could be avoided by the introduction of an adequate quality control program. The extraction step should be daily checked by training technicians as well as the introduction of routinely internal quality controls.

Other aspect of this technique that must be submitted to an adequate quality control is the deposit of the sample on the microplate and the cleaning of this microplate. It is necessary to implement a control about these issues for the correct identification of the samples.

Finally, it is a good practice for these devices to have the adequate maintenance of the MALDI-TOF based on a global quality control program. In this sense, the calibration control is a tool useful to recalibrate the spectrometer device and to identify technical problems. Appropriate maintenance of these devices is essential to carry out an adequate and accurate viral identification.

It is essential for both to continuously improve the quality of the database and to introduce a quality control program for the MS device. The implementation of these programs can help to improve the quality of the overall procedure, so it will be the basis to improve the usefulness and the accuracy of MALDI-TOF.

12.6 CONCLUDING REMARKS

MALDI-TOF MS is a relatively recent innovation for the identification of microorganisms and is being used in many clinical microbiology laboratories. The main application of this technology is, at the moment, the identification of bacteria to species level in less than a minute, although this method is about to become the standard of identification of microorganisms in most laboratories.

MALDI-TOF MS technology is also being introduced in virology assays. The main applications in this field are the identification of viruses from clinical samples, the discovery of mutations for these viruses, the rapid screening of virus subtypes, its use in molecular epidemiology, and the detection of resistance to several antiviral agents.

Future applications of MALDI-TOF MS in clinical virology will be developed to provide more accurate identification on clinical specimens and will include the detection of resistant protein markers and specific virulence for several viruses. However, the introduction of this technology routinely for virology diagnosis would enable a change in working practices. Moreover, this fact will require a complete integration into laboratory workflow and several technical improvements such as the detection of sensitivity. The continuous update of databases for microorganisms will be also an important issue, as well as the possibility for the diagnosis directly from clinical samples.

MS methods can be currently used for viral diagnosis but refinement of technology will be an important approach for the introduction of MALDI-TOF in routine viral diagnosis.

REFERENCES

[1] Anhalt JP, Fenselau C. Identification of bacteria using mass spectrometry. Anal Chem 1975;47:219−25.
[2] Claydon MA, Davey SN, Edwards-Jones V, Gordon DB. The rapid identification of intact microorganisms using mass spectrometry. Nat Biotechnol 1996;14:1584−6.
[3] Holland RD, Wilkes JG, Rafii F. Rapid identification of intact whole bacteria based on spectral patterns using matrix-assisted laser desorption/ionization with time-of-flight mass spectrometry. Rapid Commun Mass Spectrom 1996;10:12227−32.
[4] Keys CJ, Dare DJ, Sutton H. Compilation of a MALDI-TOF mass spectral database for the rapid screening and characterisation of bacteria implicated in human infectious diseases. Infect Genet Evol 2004;4:221−42.
[5] Giebel R, Worden C, Rust SM. Microbial fingerprinting using matrix-assisted laser desorption ionization time-of-flight mass spectrometry (MALDI-TOF MS) applications and challenges. Adv Appl Microbiol 2010;71:149−84.
[6] Emonet S, Shah HN, Cherkaoui A, Schrenzel J. Application and use of various mass spectrometry methods in clinical microbiology. Clin Microbiol Infect 2010;16:1604−13.
[7] Van Belkum A, Welker M, Erhard M, Chatellier S. Biomedical mass spectrometry in today's and tomorrow's clinical microbiology laboratories. J Clin Microbiol 2012;50:1513−17.
[8] Ecker DJ, Sampath R, Massire C. Ibis T5000: a universal biosensor approach for microbiology. Nat Rev Microbiol 2008;6:553−8.
[9] Stanssens P, Zabeau M, Meersseman G. High-throughput MALDI-TOF discovery of genomic sequence polymorphisms. Genome Res 2004;14:126−33.
[10] Honisch C, Chen Y, Mortimer C. Automated comparative sequence analysis by base-specific cleavage and mass spectrometry for nucleic acid-based microbial typing. Proc Natl Acad Sci USA 2007;104:10649−54.
[11] Honisch C, Chen Y, Hillenkamp F. Comparative DNA sequence analysis and typing using mass spectrometry. In: Sha HN, Gharbia SE, editors. Mass spectrometry for microbial proteomics. Chichester: John Wiley&Sons; 2010. p. 443−62.
[12] Ecker DJ, Massire C, Blyn LB. Molecular genotyping of microbes by multilocus PCR and mass spectrometry: a new tool for hospital infection control and public health surveillance. Methods Mol Biol 2009;551:71−87.
[13] Sjöholm MIL, Dillner J, Carlson J. Multiplex detection of human herpes viruses from archival specimens by using matrix-assisted laser desorption ionization-time of flight mass spectrometry. J Clin Microbiol 2008;46:540−5.
[14] Yan X, Zhong W, Tang A, Schielke EG, Hang W, Nolan JP. Multiplexed flow cytometry immunoassay for influenza virus detection and differentiation. Anal Chem 2005;77:7673−8.
[15] Townsend MB, Dawson ED, Mehlmann M. Experimental evaluation of the FluChip diagnostic microarray for influenza virus surveillance. J Clin Microbiol 2006;44:2863−71.

[16] Downard KM, Morrissey B, Schwahn AB. Mass spectrometry analysis of the influenza virus. Mass Spect Rev 2009;28:35−49.

[17] Chou TC, Hsu W, Wang CH, Chen YJ, Fang JM. Rapid and specific influenza virus detection by functionalized magnetic nanoparticles and mass spectrometry. J Nanotechnol 2011;9:52.

[18] Majchrzykiewicz-Koehorst JA, Heikens E, Trip H, Hulst AG, de Jong AL, Viveen MC, et al. Rapid and generic identification of influenza A and other respiratory viruses with mass spectrometry. J Virol Methods 2015;213:75−83.

[19] Zhang C, Xiao Y, Du J, Ren L, Wang J, Peng J, et al. Application of multiplex PCR coupled with matrix-assisted laser desorption ionization-time of flight analysis for simultaneous detection of 21 common respiratory viruses. J Clin Microbiol 2015;53:2549−54.

[20] Xiu L, Zhang C, Wu Z, Peng J. Establishment and application of a universal coronavirus screening method using MALDI-TOF mass spectrometry. Front Microbiol 2017;8:1510.

[21] Thao NTT, Ngoc NTK, Tu PV. Develoment of a multiplex polymerase chain reaction assay for simultaneous identification of human enterovirus 71 and coxsackievirus A16. J Viral Methods 2010;170:134−9.

[22] Piao J, Jiang J, Xu B. Simultaneous detection and identification of enteric viruses by PCR-mass assay. PLoS One 2012;7:e42251.

[23] Peng J, Yang F, Xiong Z. Sensitive and rapid detection of viruses associated with hand foot and mouth disease using multiplexed MALDI-TOF analysis. J Clin Virol 2013;56:170−4.

[24] Yi X, Li J, Yu S. A new PCR-based mass spectrometry system for high-risk HPV. Part I. Am J Clin Pathol 2011;136:913−19.

[25] Duu H, Yi J, Wu R. A new PCR-based mass spectrometry system for high risk HPV. Part II. Am J Clin Pathol 2011;136:920−3.

[26] Cricca M, Marasco E, Alessandrini F, Fazio C, Prossomariti A, Savini C, et al. High-throughput genotyping of high-risk human papillomavirus by MALDI-TOF mass spectrometry-based method. New Microbiol 2015;38:211−23.

[27] Ganova-Raeva L, Ramachandran S, Honisch C. Robust hepatitis B virus genotyping by mass spectrometry. J Clin Microbiol 2010;48:4161−8.

[28] Hong SP, Kim NK, Hwang SG. Detection of hepatitis B virus YMDD variants using mass spectrometry analysis of oligonucleotide fragments. J Hepatol 2004;40:837−44.

[29] Luan J, Yuan J, Li X. Multiplex detection of 60 hepatitis B virus variants by MALDI-TOF mass spectrometry. Clin Chem 2009;55:1503−9.

[30] Rybicka M, Stalke P, Bielawski KP. Dynamics of hepatitis B virus quasispecies heterogeneity in association with nucleos(t)ide analogue treatment determined by MALDI-TOF MS. Clin Microbiol Infect 2015;21:288.

[31] Kim YJ, Kim SO, Chung HJ. Population genotyping of hepatitis C virus by matrix-assisted laser desorption/ionization time-of-flight mass spectrometry analysis of short DNA fragments. Clin Chem 2005;51:1123−31.

[32] Ilina EN, Malakhova MV, Generozov EV, Nikolaev EN, Govorun VM. Matrix-assisted laser desorption ionization-time of flight (mass spectrometry) for hepatitis C virus genotyping. J Clin Microbiol 2005;43:2810−15.

[33] Oh HB, Kim SO, Cha CH. Identification of hepatitis C virus genotype 6 in Korean patients by analysis of 5' untranslated region using a matrix assisted laser desorption/ionization time of flight-based assay restriction fragment mass polymorphism. J Med Virol 2008;80:1712−19.

[34] Bayliss J, Moser R, Bowden S, McLean CA. Characterisation of single nucleotide polymorphisms in the genome of JC polyomavirus using MALDI TOF mass spectrometry. J Virol Methods 2010;164:63−7.

[35] Yea C, McCorrister S, Westmacott G, Petric M, Tellier R. Early detection of influenza A (H5) viruses with affinity for the human sialic acid receptor by MALDI-TOF mass spectrometry based mutation detection. J Virol Methods 2011;172:72—7.

[36] Zürcher S, Mooser C, Lüthi AU. Sensitive and rapid detection of ganciclovir resistance by PCR based MALDI-TOF analysis. J Clin Virol 2012;54:359—63.

[37] Seng P, Drancourt M, Gouriet F. Ongoing revolution in bacteriology: routine identification of bacteria by matrix-assisted laser desorption ionization time-of-flight mass spectrometry. Clin Infect Dis 2009;49:543—51.

[38] Cherkaoui A, Hibbs J, Emonet S. Comparison of two matrix-assisted laser desorption ionization-time of flight mass spectrometry methods with conventional phenotypic identification for routine identification of bacteria to the species level. J Clin Microbiol 2010;48:1169—75.

[39] Welker M. Proteomics for routine identification of microorganisms. Proteomics 2011;11:3143—53.

CHAPTER THIRTEEN

Identification of Mycobacteria by Matrix-Assisted Laser Desorption Ionization—Time-of-Flight Mass Spectrometry

Javier Rodriguez-Granger[1], Julián Ceballos Mendiola[1]
and Luís Aliaga Martínez[2]
[1]Department of Microbiology, Hospital Virgen de las Nieves, Granada, Spain
[2]Departament of Medicine, University of Granada, Granada, Spain

13.1 INTRODUCTION

Mycobacterium, consisting of approximately 170 species (http://www.bacterio.net/mycobacterium.html), is the only genus in the family Mycobacteriaceae and has undergone tremendous taxonomic revision in recent decades. They are aerobic (though some species are able to grow under reduced-O_2 atmosphere), non-spore-forming, nonmotile, rod-shaped bacteria which are characterized by a high Guanidine-cytosine (GC) content and complex lipid rich cell wall comprised mycolic acids which can total up to 60% of the dry weight [1]. Colony morphology varies among the species, ranging from smooth to rough and from non-pigmented to pigmented. Some species require light to form pigment (photochromogens), other species form pigment in either the light or the dark (scotochromogens). The genus *Mycobacterium* includes obligate pathogens, opportunistic pathogens, and saprophytes. Nutritional requirements to growth for most species are simple substrates, using ammonia or amino acids as nitrogen sources and glycerol as a carbon source in the presence of mineral salts. Growth of mycobacteria is stimulated by carbon dioxide and by fatty acids which may be provided in the form of egg yolk or oleic acid [2].

Four groups of human pathogens are recognized in the *Mycobacterium* genus: *Mycobacterium tuberculosis* complex, *Mycobacterium leprae*, slowly

The Use of Mass Spectrometry Technology (MALDI-TOF) in Clinical Microbiology.
DOI: https://doi.org/10.1016/B978-0-12-814451-0.00013-7

growing nontuberculous and rapidly growing mycobacteria. They were also classified by Timpe and Runyon according to their growth rate and pigment formation (Types I, II, III, and IV) [3].

Most of them are with a great clinical repercussion, because they are the causative agents of infections in humans and animals with an important morbidity and mortality. The number of infections caused by mycobacteria has increased over the past few decades. In the developed world, incidence of nontuberculous mycobacteria (NTM) is now greater than tuberculosis infection.

Two of the main pathogens of the genus, *M. tuberculosis and M. leprae* are still nowadays one of the biggest health problems in the world. In the case of *M. tuberculosis* complex, it was estimated to be responsible for 9 million cases of disease and 1.5 million deaths worldwide in 2014 [4]. In addition, the appearance and dissemination of strains resistant to multiple drugs has been witnessed in recent years [5]. Tuberculosis is caused by members of the *Mycobacterium tuberculosis* complex (MTBC), which consists of a group of closely related *Mycobacterium* species, including *M. tuberculosis* sensu stricto, *Mycobacterium bovis*, *M. bovis* bacillus Calmette–Guérin (BCG), *Mycobacterium africanum*, and four other rarely isolated members: *Mycobacterium microti*, *Mycobacterium canettii*, *M. caprae*, and *Mycobacterium pinnipedii* [6]. Species of this complex are the primary cause of tuberculosis in humans and also infect wild and domesticated animals. MTBC subspecies differ in their range of hosts, geographic distribution, drug resistance, and pathogenicity [7]. On the other hand, leprosy continues to be a public health problem, with the majority of cases focusing on Asia, America, and South America. In 2015, the number of new cases worldwide was 174,608 [8].

Many NTM are well-established pathogens and may be increasing in prevalence in part due to increased numbers of immunocompromised individuals as well as to the increasing prevalence of medical hardware and indwelling devices [9]. These bacteria have been variously described by terms "atypical mycobacteria other than tuberculosis" "environmental," "environmental opportunistic," and most commonly "nontuberculous" or "NTM."

It is recommended to carry out a precise identification at the species level in case of any isolation of mycobacteria; it is critical to distinguish pathogenic species from common environmental contaminants (such as *Mycobacterium gordonae*) and establish appropriate treatment. The most notable members of the NTM are the species and subspecies that make up the *Mycobacterium avium* complex (MAC). Within this complex, the species *M. avium* and *Mycobacterium intracellulare* are the most common NTM infection in patients with AIDS and pulmonary infections, and recently *Mycobacterium chimaera* has been implicated in endocarditis infections through contaminated water within equipment used during heart surgery. They are found in both fresh and all water sources, as well as various animals [10,11]. Another important pathogen in respiratory disease is *Mycobacterium kansasii*. The primary manifestation of *M. kansasii* infection resembles that of pulmonary tuberculosis, with cavitary infiltrates in the upper lobes. However, noncavitary and nodular bronchiectatic lung disease has also been observed. Risk factors include pneumoconiosis (especially in miners), chronic obstructive lung disease, malignancy, and alcoholism [12]. *Mycobacterium abscessus* and its subspecies are often isolated in clinical specimens from patients. These organisms are environmental, being found worldwide in water, soil, and dust, and are responsible for a wide variety of infections. These organisms often cause infections of skin and soft tissue, but they can cause more serious infections, including disseminated disease. Severe lung infections can occur in persons with underlying chronic lung disease, including patients with cystic fibrosis [13]. *Mycobacterium chelonae* complex consists of *M. chelonae* and three additional validated species (*Mycobacterium franklinii*, *Mycobacterium immunogenum*, and *Mycobacterium salmoniphilum*) [12]. The most common clinical manifestations are skin, soft tissue, and bone infections, often related to infected piercing wounds, contaminated tattoo inks, plastic surgery, or liposuction. Disseminated diseases have been described in immunocompromised individuals, especially in those receiving high-dose steroids. Infections may also be associated with ophthalmic surgery or contact lens wear (keratitis). Pulmonary infections by *M. chelonae* are less common than are those by *M. abscessus* [14].

The *Mycobacterium fortuitum* complex includes the species *M. fortuitum*, *Mycobacterium peregrinum*, *Mycobacterium senegalense, Mycobacterium setense, Mycobacterium septicum, Mycobacterium porcinum, Mycobacterium houstonense, Mycobacterium boenickei, Mycobacterium brisbanense*, and *Mycobacterium neworleansense. Mycobacterium fortuitum* is frequently associated with skin, soft tissue, and bone infections, while it rarely causes pulmonary disease, except in cases of lipoid pneumonia, gastroesophageal disorders, or disseminated diseases [12].

The phenotypic identification of mycobacteria is slow, laborious, and with little capacity to discriminate the different species described today, and classically relied on growth rate, morphological features, preferred growth temperature, pigmentation, enzymatic properties, and high-performance liquid chromatography (HPLC), which generates species or complex specific mycolic profiles [15,16]. The commercial molecular methods are the most used, being limited also to a number of species and have a high cost. DNA sequencing of 16 rRNA, *rpoB*, and *hsp65* genes is widely considered the gold standard for identification [17]. These methods are costly, can be labor-intensive, require specific equipment and expertise, and are often restricted to reference laboratories.

The use of matrix-assisted desorption ionization—time-of-flight mass spectrometry (MALDI-TOF MS) serves to detect protein profiles and perform in a few minutes the precise identification of conventional bacteria, including *Mycobacterium* spp. [18]. But in these last ones by their special characteristics, especially the bacterial wall, they make their processing and interpretation difficult. [19].Several protocols have been specifically designed for MALDI-TOF MS identification of tuberculous mycobacteria and NTM. All of these protocols include pre-MALDI-TOF MS processing of the isolate. Processing includes suspension in water and alcohol for partial inactivation, heat inactivation, and mechanical disruption with silica beads followed by acetonitrile/formic acid for mycobacterial protein extraction, spotting onto a MALDI-TOF target and cover with matrix solution (saturated α-cyano-4 hydroxycinnamic acid, 50% acetonitrile, and 2.5% trifluoroacetic acid). Each of these steps and protocols, in addition to the breadth and depth of the MALDI-TOF MS reference databases, may introduce assay variability and influence the analytical performance characteristics of the method.

13.2 STRUCTURE AND COMPOSITION CELL WALL OF MYCOBACTERIA

The *Mycobacterium* genus has different characteristics than Gram-positive and -negative bacteria, especially in reference to its cell wall. Much of the early structural definition of the cell wall was conducted in the 1960s and 1970s and later continued by Minnikin et al. who in 1982 proposed the currently accepted structural model for the cell wall architecture [20]. It contains a large amount of lipoarabinomannan in its structure, the most important being the quantity and complexity of the lipids it harbors. Among them, mycolic acids that are esterified by arabinogalactan, which is also linked to peptidoglycan, form the authentic skeleton of the mycobacterial cell wall. The hydrocarbon chains of mycolic acids are interspersed with a numerous lipids and glycolipids associated with the wall. This fact gives them characteristic properties such as acid—alcohol resistance, slow growth, resistance to common detergents and antimicrobials, and antigenicity [2].

The complexity of the bacterial wall of mycobacteria is a serious handicap when it comes to study, either in the nucleic acid detection or study of proteins.

13.3 COMMERCIAL MS PLATFORMS FOR MYCOBACTERIAL IDENTIFICATION

MALDI-TOF is being used with increasing frequency to identify mycobacterial isolates. MALDI-TOF MS detects the abundance of proteins with specific mass to charge ratios, which is displayed as a spectrum. The spectral data are then compared to a database to determine the likely identity of the organism. Currently two MALDI-TOF have been commercialized, the Bruker Microflex LT Biotyper system (Bruker Daltonics, Bremen, Germany) and the VITEK MS system (BioMérieux, Marcy l' Etoile, France). Although in essence they are quite similar, they present some differences in terms of the database and the interpretation of the results. Due to the difficulties encountered at first, the databases are still in development and new versions appear that incorporate more species and, therefore, improve the accuracy of the identification. Similar to rRNA sequencing, the accuracy of MALDI-TOF is dependent on both the robustness of the database and quality of spectra obtained. In the case of identification of mycobacteria, due to their inherent difficulties, a decrease score value to 1.8 for high confident species and 1.6 low confident species identification is generally accepted by most laboratories [21]. Manual acquisition appears to improve the quality of the spectra and therefore increases species identification, particularly at the lower score. Factors such as their robust cell wall, dense mycolic acid layers, and relatively low concentration of ribosomal proteins compared to other microorganisms exacerbate the problem of obtaining high-confidence identification scores.

The VITEK MS system has an FDA-cleared database which includes mycobacteria [22]. Laboratories using the MALDI Biotyper must build their own databases or rely on research using only databases for their laboratory-developed protocols [23]. The MALDI Biotyper system has a database called "Mycobacteria library v 2" that contains 313 entries. In their development, strains of culture collection and clinical isolates were used. This database allows the identification of 131 species, but really only identifies 127, since the remaining 4 are combined in different groups. A new version "Mycobacteria library v 3.0" has recently appeared with a total of 853 entries, adding 21 missing species in the previous version 2.0 and eliminating 3 that were present, becoming able to identify a total of 149 species. Finally, version Mycobacteria v 4.0 has been presented, which has 880 entries of which 450 are clinical isolates that include 159 species [24].

The databases present other limitations and characteristics among which the following should be mentioned: in the tuberculosis complex it does not differentiate all the species, although it correctly identifies the complex as such. *M. abscessus* requires small modifications to differentiate the three subspecies. In the case of MAC it is possible to differentiate subspecies with *M. avium* spp. avium, *M. avium* spp. paratuberculosis, and *M. avium* spp. silvaticum. Database VITEK IVD v 3.0 can separate *M. intracellulare* and *M chimaera*, although the similarity is so great that the results are inconclusive [22].

13.4 IDENTIFICATION OF MYCOBACTERIA BY MS

The MS for mycobacteria as we have mentioned previously has different aspects that hinder and affect the results obtained through this technology.

13.4.1 Inactivation of microorganisms

Several inactivation protocols have been developed to minimize the risk of exposure of laboratory personnel to MTBC and to maximize the quality of the spectra obtained. The most used: the treatment by heat, chemical agents, radiations and the combination of these. In a few studies, mycobacteria were analyzed without previous inactivation

[25,26] which does not comply with current standards for the manipulation of harmful organism. MTBC are Biosafety Level 3 organisms that must be inactivated prior to manipulation outside a biological safety cabinet to avoid potentially exposing the laboratory staff to contamination [2]. There is no standard inactivation procedure currently available for identifying mycobacteria by using MALDI-TOF for either the Bruker or bioMérieux system. Previously published procedures for mycobacterial protein extraction have either been technically complex, using materials not often found in a clinical microbiology laboratory, or have involved multiple centrifugation and washing steps prior to organism inactivation [27].

There are different works published in the last years that have evaluated different inactivation protocols, the study published by El Khéchine et al. in 2011 uses two inactivation protocols (95°C/60 minutes and 70% de ethanol/10 minutes) not finding growth of *M. tuberculosis*, *M. bovis* BCG, and *M. fortuitum* over 45-day incubation after either inactivation protocol used. However, the incubation of Middlebrook 7H9-grown mycobacteria in ethanol yielded protein profiles of lower quality than those obtained after heat inactivation. Ethanol has a protein precipitating action. Other agents used that can influence the quality of the spectra obtained include the use of glutaraldehyde 2% [28]. In another paper published by Machen et al. in 2013, two protocols were described, Protocol A: using heat inactivation in ethanol followed by sonication (30 minutes at 95°C), and Protocol B: using cell disruption with glass bead in ethanol. After inactivation by Protocol A, 88/107(82.2%) mycobacterial isolates were correctly identified to the species or genus level. After inactivation by Protocol B, 95/107 (88.8%) mycobacterial isolates were identified to the species or genus level [29]. In the last paper published by Mather et al. in 2014, two simplified protein extraction protocols developed at University of Washington (UW) and by bioMérieux (BMX) for use with two different MS platforms were tested. Both protocols included vortexing with silica beads in the presence of ethanol. Only the UM protocol makes an inactivation to heat at 95°C for 30 minutes. To assess the efficacy of inactivation, the samples were inoculated on Middlebrook 7H11 agar and into VersaTREK Myco Bottles; no growth was observed on either solid medium or in liquid broth for any of the organisms tested after 6 weeks of incubation. Both procedures produced high-quality spectra with an average of greater than 50 peaks per spectrum [18].

As we have seen, the different methods of inactivation stated allow us to work in a safe way because they achieve a complete inactivation of the microorganisms.

13.4.2 Methods of sample preparation

There is no final agreement on the use of total bacteria or the use of extracts of these for mass spectroscopy analysis. However, with the extraction protocols suggested by MALDI Biotyper and VITEK MS, the overall performance of MS has not been optimal. Multiple modifications have been proposed and analyzed in the preparation of the samples to be analyzed. It was first reported by Hettick et al. that the analysis of mycobacterial whole cells by MALDI-TOF MS could be used for identification. It was demonstrated that analysis of acetonitrile/trifluoroacetic acid cellular extracts produces data similar to that of the analysis of deposited whole cells, while minimizing human contact with the microorganisms and rendering them nonviable. A matrix composition of alpha-cyano-4-hydroxycinnamic acid with fructose yields highly reproducible MALDI-TOF spectra [26].

Pignone et al. in 2006 show that identification of diverse species of *Mycobacterium* and strains within those species is possible using whole-cell MALDI. All of the 37 strains, representing 13 species belonging to four groups of *Mycobacterium* were analyzed yielding reproducible unique mass spectral profiles [25]. The techniques used in both publications require several steps: Pignone et al. describe a strategy based on root mean square values to compare different profiles. Hettick et al. developed a biostatistical analysis to identify mycobacterial species.

A limitation of these studies is that the number of tested strains is less than 40, encompassing a maximum of 13 species. In a short time other articles were published with a greater number of strains, with the paper published by Lozt et al. in 2010 where a total of 311 strains belonging to 31 distinct species and 4 species complexes grown in Lowenstein—Jensen (LJ) and liquid (mycobacterium growth indicator tube [MGIT]) media were analyzed. No extraction step was required. Correct identifications were obtained for 97% of strains from LJ and 77% from MGIT media. No misidentification was noted. Also they showed that a high number of replicates increases the probability of good identification, especially for slow-growing mycobacteria: in some cases, five replicates were required to obtain one good spectral acquisition [30]. This is a striking difference from the identification of routinely isolated bacteria by MALDI-TOF MS, which usually requires only one replicate. Briefly, the 70% ethanol suspension was deposited in five replicates on a target plate and allowed to dry at room temperature. On each well, 1 μL of matrix solution synapinic acid (SA) was added and allowed to crystallize with the sample at room temperature. One microliter of 10 mM phosphate was then added to each well. A number of important recently published articles describe different extraction protocols, as in the case of the article published by Khéchine et al. in 2011 [28] where eight different extraction protocols were examined, with variations in the inactivation process (95°C/60 minutes or 70% ethanol/10 minutes) and in the type and composition of the reagents used to solubilize, wash, and precipitate mycobacterial proteins. With the intention of decrease the inoculum required for the accurate MALDI-TOF MS identification of mycobacteria. The final protocol consisted of different steps : colonies from each isolate were collected in an Eppendorf tube containing 500 μL of HPLC-grade water and 0.5% Tween-20 and inactivated by heating at 95°C for 1 hour. The mycobacterial suspension was then washed twice with 500 μL of HPLC-grade water and centrifuged at 13,000 $\times$ g for 10 minutes. The pellet was vortexed along with 500 mL of HPLC grade water and 0.3 g acid-washed glass beads (diameter <106 mm, Sigma) in a BIO 101 FastPrep instrument (Qbiogene, Strasbourg, France) at level 6.5 (full speed) for 3 minutes. The suspension was centrifuged at 13,000 $\times$ g for 10 minutes. The pellet was then resuspended in 5−50 mL 70% HCOOH and 5−50 mL 100% acetonitrile, depending on the volume of the pellet. The suspension was centrifuged at 11,000 $\times$ g for 1 minute and 1.5 mL of the supernatant was deposited on a target plate (Bruker Daltonics, Bremen, Germany) in four replicates. Finally, 1.5 mL matrix solution (saturated a cyano-4-hydroxycinnamic acid, 50% acetonitrile, 2.5% trifluoroacetic acid) was added and allowed to co-crystallize with the sample pellet at room temperature.

The inactivation protocols that include heating are very effective but acid-fast staining of the heated samples showed that the mycobacteria had clumped together. Different methods for dispersing the clumps were compared, including repeatedly passing the suspension through a narrow gauge needle into a syringe, grinding the specimen in a 15-mL conical tube with a tissue grinder, and grinding the specimen in a 1.5-mL Eppendorf tube with a micropestle. Acid-fast staining demonstrated that the micropestle had dispersed the clumps more than the other techniques.

Despite the successful disruption of clumps with the micropestle, the quality of spectra for several species was still poor and precluded adequate organism identification.

Another problem that the extraction protocols present is the presence of glycopeptidolipids in the cell wall, as has been shown in the case of the MAC and *Mycobacterium smegmatis* that prevent the entry of extraction reagents [31].

The use of silica balls improved the dispersion of the mycobacterial suspension, which together with the use of different agitators allows to apply diverse and intense movements that favor the rupture and fragmentation of the mycobacteria achieving higher quality of protein spectra [27,32].

Also the use of silica beads with formic acid for protein extraction the possibility was considered that heat generated during vortexing may lead to formylation of the proteins and subsequently result in misidentification of the organisms. No formylation occurred in the samples that were vortexed for 10 or 20 minutes, and a minimal amount of formylation occurred in the sample that was vortexed for 60 minutues. Consequently, no changes were needed in 10-minute extraction protocol [27].

Recently an improvement MALDI TOF MS identification of *Mycobacterium* spp. has been described by the use of a novel two-step cell disruption preparatory technique, proposing that the combined use of 0.5 mm diameter silica/zirconia beads and sonication for 15 minutes produced greatly improved results [33].

13.4.3 Growth from solid and liquid cultures of *Mycobacterium* spp.

A combination of broth and solid media should be included for inoculation. Broth medium is essential for the recovery of *Mycobacterium* spp. in as short an incubation time as possible. There are many types of broth media, many of which are included with automated mycobacterial detection systems; most are based on Middlebrook 7H9 broth. Solid media (egg based and non−egg based) are included to ensure the recovery of rare strains that may not grow in broth.

The influence of cultivation medium or length of cultivation on the quality of MALDI-TOF MS profiles has been reported for different bacterial species [34].

A great variability between solid and liquid culture media has been observed in *Mycobacteria* genus for use in MALDI-TOF MS, the only study on this topic carried by Lozt et al. in 2010, who cultivated myco-bacteria in both solid and liquid medium demonstrating that the protein extract obtained from bacterial cells cultivated in solid medium provided mass spectra of better quality and higher identification success rate. This is mostly due to spectral acquisition failures, due either to the low number of bacteria or to potential interference of the supplements included in liq-uid medium [30]. On the other hand, liquid media—through the auto-matic incubation and detection systems BACTEC MIGIT 960 (BD Diagnostics), BactT/ALERT 3D (bioMérieux), and VersaTREK (Thermo Scientific)—are more sensitive than the media solids. This greater sensitivity implies that the growth is detected with a lower num-ber of microorganisms. The smallest inoculum, obtained by using the liq-uid medium, would entail a decrease in the sediment or pellet necessary for optimal extraction required in the mass spcetrometry (MS) analysis.

Concerning the length of cultivation, Balazova et al. in 2014 did not observe any significant influence on the automatic acquisition of spectra and the number of repeatable peaks. While identification of the genus level was not notably influenced by cultivation time, for species identifica-tion, sample preparation after 3 days of cultivation seemed to be prefera-ble [34]; however, other authors describe finding better results with the most grown cultures [30].

13.4.4 Reading and interpretation of protein profiles

Obtaining correct identification through MS is conditioned, in the first place, to obtain a good protein profile and, secondly, to the quality of the database with which it must be compared. The reading of protein profiles in different platforms available is automatic. However, the automatic read-ing does not always get the number of enough peaks so that they can be properly evaluated by the analysis programs. This may be due to various factors such as extraction protocols or the conditions and type of culture. For this reason, in daily practice we must resort to reading and manual analysis of MS.

Another important aspect to consider is the number of replicates of the samples to be studied. The higher the replicas to be studied, the greater the probability of obtaining good protein profiles and thus obtain-ing a good identification. The degree of correlation between MS and standard methods in the identification of the genus *Mycobacterium* shows some differences by analyzing separately the MTBC and MNT.

With respect to the first, the correlation between MS and the standard methods has been good and has not generated any problem of error in the identification of the complex [32,35]. However, this is not the case among the various species that make it up, and the differences in the differentiation between M. *tuberculosis* and M. *bovis*—including BCG—are particularly relevant because of the clinical and therapeutic impact that this entails [32].

In the case of NMT the greatest disagreements with the reference methods have been detected in phylogenetically related species. Likely, some of the differences obtained can be explained by the use of incomplete or non-updated databases. Despite this, recently Rodríguez-Sánchez et al. in 2015 make a comparison by 125 NMT analyzed by MALDI-TOF and Genotype CM/AS, considered 16S rRNA/hsp 65 sequencing the gold standard, finding agreements with reference method of 94.4% and 84%, respectively [36]. Another study published in the same years finds correlation between MALDI-TOF and the standard methods of 63.8% (155/243). According to the type of culture medium in which the strains were isolated, the correlation was 71.3% (82/115) in solid medium and 57.8% (74/128) in liquid medium [35].

Several studies have demonstrated the ability to identify mycobacteria including M. *avium*, M. *immunogenum*, M. *chelonae*, and M. *abscessus*. However, the subspecies of M. *abscessus*, M. *avium*—M. *intracellulare*, or rapid growth NMT cannot be differentiated by MALDI-TOF [18,27,30]. MALDI-TOF MS distinguished two M *chimaera*—M *intracellulare* groups separated from M. *avium* and from the other mycobacterial species on as core-oriented dendrogram, but it also failed to differentiate the two species [37]. Misidentification and lack of correlation with standard methods has mainly been reported in phylogenetically close species [27,35].

13.5 CONCLUSION

MALDI-TOF MS has recently been reported to be a reliable and expedited method for identification of mycobacteria. This technique is rapid, providing results within a few hours, being faster than sequencing and hybridization-based, and reduce identification-related cost. However, it is necessary for future improvements in database and extraction protocol to achieve a higher rate of identification mainly when it is made from liquid media.

REFERENCES

[1] Madigan M. Brock biology of microorganisms. 13th ed. International Microbiology; 2012.

[2] Pfyffer G. Mycobacterium: general characteristics, laboratory detection, and staining procedures. In: Jorgensen J, Pfaller MA, Carrol KC, Funke G, Landry ML, et al., editors. Manual of clinical microbiology. 11th ed. Washington, DC: ASM Press; 2015.

[3] Timpe A, Runyon EH. The relationship of atypical acid—fast bacteria to human disease: a preliminary report. J Lab Clin Med 1954;44:202—9.

[4] World Health Organization. Global tuberculosis report 2014. Geneva, Switzerland: WHO; 2014.

[5] World Health Organization. Global tuberculosis report 2016. Geneva, Switzerland: WHO; 2016.

[6] 2nd ed. Goodfellow M, Kämpfer P, Busse H-J, Trujillo ME, Suzuki K, Ludwig W, et al., editors. Bergey's manual® of systematic bacteriology. The Actinobacteria, Part A and B, Vol. 5. New York, NY: Springer-Verlag; 2012. pp. XLVIII—2083.

[7] Wayne L, Kubica G. The mycobacteria. In: Sneath PHA, Mair NS, Sharpe ME, Holt JG, editors. Bergey's manual of systematic bacteriology, vol. 2. Baltimore, MD: Williams & Wilkins; 1986. p. 1435—57.

[8] World Health Organization. Global leprosy update, 2015: time for action, accountability and inclusion. Wkly Epidemiol Rec 2016;91:405—20.

[9] Joint Tuberculosis Committee of the British Thoracic Society. Management of opportunist mycobacterial infections: Joint Tuberculosis Committee Guidelines 1999. Subcommittee of the Joint tuberculosis. Committee of the British Thoracic Society. Thorax 2000;55:210—18.

[10] Bryant JM, Grogono DM, Greaves D, Foweraker J, Roddick I, Inns T, et al. Whole-genome sequencing to identify transmission of *Mycobacterium abscessus* between patients with cystic fibrosis: a retrospective cohort study. Lancet 2013;381:1551—60.

[11] Haller S, Holler C, Jacobshagen A, Hamouda O, Abu Sin M, Monnet DL, et al. Contamination during production of heater-cooler units by *Mycobacterium chimaera* potential cause for invasive cardiovascular infections: results of an outbreak investigation in Germany, April 2015 to February 2016. Euro Surveill 2016;21(17) pii_30215.

[12] Griffith DE, Aksamit T, Brown-Elliott BA, Catanzaro A, Daley C, Gordin F, et al. ATS Mycobacterial Diseases Subcommittee, American Thoracic Society, Infectious Disease Society of America An official ATS/IDSA statement: diagnosis, treatment, and prevention of nontuberculous mycobacterial diseases. Am J Respir Crit Care Med 2007;175:367—416.

[13] Centers for Disease Control and Prevention. *Mycobacterium abscessus* in healthcare settings. Atlanta, GA: Centers for Disease Control and Prevention; 2010.

[14] Umrao J, Singh D, Zia A, Saxena S, Sarsaiya S, Singh S, et al. Prevalence and species spectrum of both pulmonary and extrapulmonary nontuberculous mycobacteria isolates at a tertiary care center. Int J Mycobacteriol 2016;5:288—93.

[15] Wayne LG, Engel HWB, Grassi C, Gross W, Hawkins J, Jenkins PA, et al. Highly reproducible techniques for use in systematic bacteriology in the genus *Mycobacterium*: tests for niacin and catalase and for resistance to isoniazid, thiophene 2-carboxylic acid hydrazide, hydroxylamine, and *p*-nitrobenzoate. Int J Syst Evol Microbiol 1976;26:311—18.

[16] Butler WR, Guthertz LS. Mycolic acid analysis by high-performance liquid chromatography for identification of *Mycobacterium* species. Clin Microbiol Rev 2001;14:704—26.

[17] Yam WC, Yuen KY, Kam SY, Yiu LS, Chan KS, Leung CC, et al. Diagnostic application of genotypic identification of mycobacteria. J Med Microbiol 2006;55:529−36.

[18] Mather CA, Rivera SF, Butler-Wu SM. Comparison of the Bruker Biotyper and Vitek MS matrix-assisted laser desorption ionization-time of flight mass spectrometry systems for identification of mycobacteria using simplified protein extraction protocols. J Clin Microbiol 2014;52:130−8.

[19] Chen JH, Yam WC, Ngan AH, Fung AM, Woo WL, Yan MK, et al. Advantages of using matrix-assisted laser desorption ionization-time of flight mass spectrometry as a rapid diagnostic tool for identification of yeasts and mycobacteria in the clinical microbiological laboratory. J Clin Microbiol 2013;51:3981−7.

[20] Minnikin DE. Lipids: complex lipids, their chemistry, biosynthesis and roles. In: Ratledge C, Stanford J, editors. The biology of the mycobacteria. London: Academic Press; 1982. p. 95−184.

[21] Rose G, CulaK R, Chambers T, Gharbia S, Shan N. The challenges of identifying mycobacterium to the species level using MALDI-TOF MS. MALDI-TOF and tandem MS for clinical microbiology. 1st ed. Chichester, West Sussex: : John Wiley Son Ltd; 2017.

[22] bioMérieux. bioMérieux receives FDA clearance for expanded pathogen identification capability on VITEK MS. Marcy l'Étoile, France: bioMérieux; 2017.

[23] Buckwalter SP, Olson SL, Connelly BJ, Lucas BC, Rodning AA, Walchak RC, et al. Evaluation of matrix assisted laser desorption ionization−time of flight mass spectrometry for identification of *Mycobacterium* species, *Nocardia* species, and other aerobic actinomycetes. J Clin Microbiol 2016;54:376−84.

[24] Bruker Daltonics, Inc. Standard operating procedure: Mycobacteria extraction (MycoEX) method (version 30). Bremen: Bruker Daltonics Inc; 2014. Available at: http://www.bruker.com.

[25] Pignone M, Greth KM, Cooper J, Emerson D, Tang J. Identification of mycobacteria by matrix-assisted laser desorption ionization-time-of-flight mass spectrometry. J Clin Microbiol 2006;44:1963−70.

[26] Hettick JM, Kashon ML, Simpson JP, Siegel PD, Maruzek GH, et al. Proteomic profiling of intact mycobacteria by matrix-assisted laser desorption/ ionization time-of-flight mass spectrometry. Anal Chem 2004;76:5769−76.

[27] Saleeb PG, Drake SK, Murray PR, Zelazny AM. Identification of mycobacteria in solid-culture media by matrix-assisted laser desorption ionization time of flight mass spectrometry. J Clin Microbiol 2011;49:1790−4.

[28] El Khéchine A, Couderc C, Flaudrops C, Raoult D, Drancourt M. Matrix-assisted laser desorption/ionization time of flight mass spectrometry identification of mycobacteria in routine clinical practice. PLoS One 2011;6:e24720.

[29] Machen A, Kobayashi M, Conelli MR, Wang YF. Comparison of heat inactivation and cell disruption protocols for identification of mycobacteria from solid culture media by use of Vitek matrix-assisted laser desorption ionization-time of flight mass spectrometry. J Clin Microbiol 2013;51:4226−9.

[30] Lotz A, Ferrori A, Beretti JL, Dauphin B, Carbonnelle E, Guet-Revillet H, et al. Rapid identification of mycobacterial whole cells in solid and liquid culture media by matrix-assisted laser desorption ionization-time of flight mass spectrometry. J Clin Microbiol 2010;48:4481−6.

[31] Schorey JS, Sweet L. The mycobacterial glycopeptidolipids: structure, function, and their role in pathogenesis. Glycobiology 2008;18:832−41.

[32] Saleeb PG, Drake SK, Murray PR, Zelazny AM. Identification of mycobacteria in solid-culture media by matrix-assisted laser desorption ionization-time of flight mass spectrometry. J Clin Microbiol 2013;49:1790−4.

[33] O'Connor JA, Lynch-Healy M, Corcoran D, O'Reilly B, O' Mahony J, Lucey B. Improved matrix assisted laser desorption ionization time of flight mass spectrometry (MALDI-TOF MS)- based identification of *Mycobacterium* spp. by use of a novel two spell disruption preparatory technique. J Clin Microbiol 2016;54:495−6.

[34] Balážová T, Makovcová J, Šedo O, Slaný M, FaldynaM, Zdráhal Z. The influence of culture conditions on the identification of *Mycobacterium* species by MALDI-TOF MS profiling. FEMS Microbiol Lett 2014;353:77−84.

[35] Tudó G, Monté MR, Vergara A, López A, Hurtado JC, Ferrer-Navarro M, et al. Implementation of MALDI-TOF MS technology for the identification of clinical isolates of *Mycobacterium* spp. in mycobacterial diagnosis. Eur J Clin Microbiol Infect Dis 2015;34:1527−32.

[36] Rodríguez-Sánchez B, Ruiz-Serrano MJ, Marín M, López-Roa P, Rodríguez-Creixems M, Bouza E. Evaluation of matrix-assisted laser desorption ionization-time of flight mass spectrometry for identification of nontuberculous mycobacteria from clinical isolates. J Clin Microbiol 2015;53:2737−40.

[37] Lecorche E, Haenn S, Mougari F, Kumanski S, Veziris N, Benmansour H, et al. Comparison of methods available for identification of Mycobacterium chimaera. Clin Microbiol Infect 2018;24(4):409−13.

CHAPTER FOURTEEN

Use of MALDI-TOF Mass Spectrometry in Fungal Diagnosis

Francisco Franco-Álvarez de Luna
Microbiology Unit, UGC, Clinical Laboratory, Hospital de Riotinto, Huelva, Spain

14.1 ORIGINS AND INTRODUCTION TO MASS SPECTROMETRY IN MYCOLOGY

New diagnostic techniques in the recent history of modern microbiology, such as the techniques of enzyme immunoassay in the early 1970s, or molecular microbiology techniques (polymerase chain reaction; PCR), in the last decades, have implied a real advantage in the day to day of the laboratories of clinical microbiology. These techniques have improved significantly on their diagnostic capacity, speed of results, as well as sensitivity and specificity. In addition, these have been able to integrate into automatic systems, and the current progressive increase of the routine workflow [1].

Similarly to the techniques previously mentioned, the introduction of matrix-assisted laser desorption/ionization—time of flight mass spectrometry (MALDI-TOF MS) has been, without any doubt, the most important technological progress in clinical microbiology during the last decade. In an only few years, it has become a broadly used technology for the daily clinical activity, available in the microbiology services of several hospital centers.

In 2009, a key article to publicize to the specialists involved in the diagnosis of infectious diseases, the possibilities of MALDI-TOF MS, was published [2]. In these studies more than 1600 isolates, including Gram-positive and Gram-negative, aerobic, and anaerobic bacteria, were analyzed, obtaining a 95.4% of correct identification.

The Use of Mass Spectrometry Technology (MALDI-TOF) in Clinical Microbiology.
DOI: https://doi.org/10.1016/B978-0-12-814451-0.00014-9

On the other hand, severe fungal infections often affecting particularly sensitive patients, such as immunosuppressed and critically ill patients, require the early establishment of an adequate antifungal therapy, thus a rapid diagnosis is fundamental for a good prognosis.

The identification of isolated organisms (such as yeasts) has been traditionally carried out through enzymatic, biochemical, and morphological studies or through fundamentally morphological studies in the case of filamentous fungi [3]. In general, these procedures require subcultures or biochemical tests that delay the microbiological result in more than 24 hours from the isolation of the pathogen. The speed, safety, and simplicity of the procedures offered by MALDI-TOF MS in microbial identification make it the ideal tool in the context of the clinical laboratory [4].

In recent years, numerous studies have demonstrated the advantages of MALDI-TOF MS for fungal identification [5,6]. In fact, its limitations tend to be more related to the availability of a complete database, rather than the ability of the method to obtain reliable profiles of almost any organism [7].

In this chapter, we will review key aspects of MS in the identification of the main clinical species, as well as other applications of great interest in the field of mycology and its potential uses in epidemiological studies or related to antifungal sensitivity.

14.2 MS YEAST IDENTIFICATION

A rapid identification of the organisms that produce fungal infections in critically ill and immunosuppressed patients is crucial. One of the main advantages of MS with greater relevance for human pathology is the accurate and rapid results in yeast identification. The emergence of MALDI-TOF MS technology incorporated into commercial systems for the use in microbial identification has been broadly used in clinical diagnostic and clinical research, which confirms this utility [8,9].

There are two main commercialized systems: MALDI Biotyper (Bruker Daltonics, Germany) and VITEK MS (BioMérieux). These offer very good results, although they vary in the method of protein extraction used, the reference library of spectra used, and the threshold chosen to consider an identification as correct. In addition, there are other aspects to be taken into account, such as the growth conditions of the strains, the incubation time, or the culture media used.

As for the extraction method, this seems to be fundamental when it comes to obtaining good results. For most bacterial species and some yeasts, the *direct method* [10] is sufficient, which consists of transferring a small amount of microorganism from the agar plate to the MALDI plate and covering them with a small amount (0.5−1 µL) of matrix solution. The solvents (50% ACN and 0.1−2.5% TFA) normally used in the matrix (CHCA; α-cyano-4-hydroxycinnamic acid, dihydroxy benzoic acid, or sinapinic acid) are sufficient to lyse the cells.

However, most fungal cells (similarly to other bacterial species such as mycobacteria) are not necessarily lysed in the same way. Different manufacturers recommend different extraction systems. Bader et al. [11] showed how the same yeast species (*Candida glabrata*) may present in different protein profiles depending on the extraction protocol.

Protein extraction method is a critical step, especially in the Bruker system, because some studies have shown that the direct extension of the colony on the plate without pretreatment only reaches to identify 10% of the isolates [12].

In addition to the direct method, there are other protocols that provide quality results: the *simplified extraction method* (VITEK MS) consisting on adding 0.5−1 µL of formic acid (25%−70%) over the extended colony in the MALDI-TOF plate before adding the matrix. And the *complete extraction method*, including a first step of inactivating the yeast with ethanol and a second step of cell lysis by means of 70% formic acid followed by acetonitrile, centrifuges to collect the supernatant. Although it is somewhat more laborious, it is the one that obtains the best results with the Bruker system [11,13].

Another important aspect to determine the performance of the MS is the reference library. Many studies confirm that the percent of correct identifications increases when these libraries are completed individually with new spectra [14]. The elaboration and validation of these databases are complex and not always available to all users. For this reason, and also for the purposes of standardization, the use of manufacturers' databases which should be suitably expanded is more appropriate [15].

On the other hand, each commercial system uses different thresholds that inform of the degree of success in the identification obtained. Some authors point out that scores of 1.8, 1.7, and even 1.5 could be accepted for the identification of some yeast species if certain criteria are met [16–18].

Pence et al. [12] have recently evaluated the growth conditions of yeasts in usual media (blood agar, BHI agar, chromogenic media, or Sabouraud with chloramphenicol), which are used by microbiology laboratories, and it does not seem to have a significant impact on the analysis. They also checked how they affected the incubation times, and how the prolongation of 24–48 hours could worsen the identification results with Bruker system, using simplified or direct extraction protocols.

Most of the studies show how MS systems obtain better results in the identification of the *Candida* genus, especially the species which are most frequently involved in the clinic. The identification of yeasts has shown remarkable results, differentiating species which are complicated to discriminate by conventional methods, such as the *Candida parapsilosis* complex [9,19].

Bader et al. [19] compared two commercial platforms (MALDI Biotyper and Axim-SARAMIS) for the identification of pathogenic yeasts. A total of 1192 isolates were analyzed, which included 36 different species. The percentage of correct identifications obtained with MALDI Biotyper platform was 97.6%, and in the case of the Axima-SARAMIS platform, the percentage was 96.1%.

Bizzini et al., in a study with 1371 isolates, and Van Veen et al., with 327 isolates, performing a previous extraction procedure, correctly identified 100% and 95%, respectively, of the yeasts isolated from various clinical samples [20,21].

MALDI-TOF has also become a powerful tool for identifying less conventional yeasts. Kolecka et al. [22] were able to identify at species level up to 98% of the 219 species studied of the genera *Galactomyces, Saprochaete, Geotrichum, Magnusiomyces, Trichosporon, and Guehomyces* spp. Other authors have also confirmed the usefulness of MS for the identification of black yeasts of the genus *Exophiala* isolated in clinical samples [23]. Subsequent studies on a larger number of isolates offer similar data, with correct identification percentages above 95% and good discrimination between species that are difficult to differentiate by other methods, such as *C. glabrata and Candida bracarensis* [19]. Only one recent study reports somewhat worse results, because among 303 yeast clinical isolates, 20 (6.6%) were not identified by MS due to deficiencies in the database; of 26 identifications discrepant with Vitek-2, in 21 of them (6.9%) it was considered that identification by MS had been erroneous when tested by genetic methods [24].

14.3 MS FILAMENTOUS FUNGI IDENTIFICATION

It is well known how a fast and reliable identification of the filamentous fungi species involved in the invasive fungal infection is fundamental for the establishment of an adequate and efficient treatment. The diagnosis of invasive fungal infection caused by filamentous fungi has become more complicated in recent years, due to the emergence of species such as *Aspergillus*, *Fusarium*, or *Scedosporium*, and the increase in the number of immunosuppressed patients susceptible to suffering such mycosis.

MALDI-TOF MS has supposed a revolution in mycology laboratories, and among its main advantages, the decrease of the response time stands out, reducing the hours or days that were necessary to carry out a genotypic or phenotypic identification through the conventional methodology, to the short 10—15 minutes by MS. On the other hand, its greater power of discrimination, precision, or safety with respect to the classic methods of filamentous fungi identification also stands out, as well as the possibility of differentiating closely related species that are morphologically very similar but with radically different antifungal sensitivity profiles [25].

However, two fundamental issues, such as the standardization of extraction methods or the expansion of commercial databases, have yet to be improved, due to the enormous biological complexity of filamentous fungi.

Regarding the extraction methodology, the lysis of the fungal cells has an enormous influence on the fungal profiles of MS, due to factors such as the thickness of the cell wall, the different stages of maturation of the selected colonies, or the presence or simultaneous absence of hyphae and conidia. The extraction protocol and the chosen matrix can introduce a greater degree of variation in the number and identity of the obtained mass spectra [26].

Bruker Daltonics [27], whose database of filamentous fungi, on its MBT 6903 MSP Library version, includes 25 genera and 42 species, recommends rapid culture in a liquid medium overnight, followed by centrifugation, washing, and extraction with ethanol, formic acid, and acetonitrile. Bruker uses liquid cultures of filamentous fungi in Sabouraud for the construction of a *"fungal library"* in an effort to minimize the effect of culture and growing conditions to help the production of a uniform mycelium. Nevertheless, the establishment of protocols that require subcultures or longer incubation times would significantly affect one of the main advantages of this technology, which is the possibility of quickly identifying from the primary culture plate. For its part, MS–VITEK, in its version 3.0, includes 32 genera and 81 species. BioMérieux does not recommend culture in liquid medium but a conventional extraction with ethanol, formic acid, and acetonitrile.

Several authors have attempted to standardize growth conditions and sample preparation for the identification of filamentous fungi to build *fungal profiles* to compare with the MALDI-TOF MS reference libraries. Cassagne et al. [28] developed a standard procedure consisting of a chemical extraction of the filamentous fungal colonies, cultured in Sabouraud medium, with formic acid and matrix of CHCA. The authors achieved the correct identification of 87% of the filamentous fungi species. On the other hand, Alanio et al. [29] developed a simple and rapid protocol that consists of depositing the superficial material (a mixture in water of the spores and the mycelium, conidiophores collected from the surface of the colonies of the filamentous fungi) directly on the MALDI-TOF plate without subculture or preparing the colony. The result was a correct identification in 98.6% (138 of 140 *Aspergillus* isolates) with 100% specificity. Lau et al. [30] developed a database of filamentous fungi, cultured on solid media with a colony less than 5 mm in diameter (first stages of mycelial growth). Its fungal library with 294 fungal isolates from 76 genera and 152 species allowed accurate identification at species level (score ≥ 2.0) and genus level (score ≥ 1.7) at 88.9% and 4.3% of isolates, respectively.

Another problem that arises is the existence of significant differences in protein profiles obtained depending on the age of the cultures, and even among different subcultures of the same strain [31]. Thus, the elaboration of extensive and complex databases, including higher number of strains and cultures profiles of different ages is required [32].

Considering all the published studies, until the date, it has been reported that an accurate identification depends on the creation of a database that contains several protein profiles for each species, eventually improving the architecture of the commercial reference libraries. Studies show that a well-prepared and sufficiently complex database is essential [28]. However, the elaboration and validation of these databases is not available to many users. So, for the purpose of standardization, the use of manufacturers databases is recommended, which should be suitably improved and expanded. Currently, the identification of filamentous fungi by MALDI-TOF MS cannot yet completely replace the conventional identification methodology [15,28,31,33,34].

14.4 MS DERMATOPHYTES IDENTIFICATION

Dermatophyte fungi (DF) are responsible for a wide variety of infections of the skin, nails, and hair. Their identification remains particularly challenging due to their morphology, both macroscopic and microscopic, which requires a long response time and professionals with extensive experience. Despite these difficulties, identification at species level is always advisable from the epidemiology and treatment [35]. The sequencing techniques are, at the moment, the gold standard for DF identification [36]. However, species like *Trichophyton soudanense* are indistinguishable by sequencing *of Trichophyton rubrum.* Considering that conventional identification delays diagnoses considerably, and that if current sequencing criteria are used, some information may be lost; it is necessary to look for other types of solutions that MS can provide.

Recent studies have shown that the accuracy of identification based on MALDI-TOF MS may vary between 13%, 5%, and 100% for dermatophytes. This variability was due to both the preanalytical phase of the process and the reference spectra library.

In general, MALDI-TOF MS provides a high level of concordance with the sequencing, although they present some limitations, because they do not use standardized protocols neither in the incubation time of the colonies or in the extraction systems used. Packeu et al. [36] obtain better results in cultures of more than 14 days (100% of correct identifications), while the percentage decreases with culture of 7 days (80%) and especially above all with those of 3 days (33%—40%). Most authors use protein extraction [37] because direct analysis, with or without formic acid, do not give adequate results. Nevertheless, the latter procedure is not standardized and the most commonly used is the one by Cassagne et al. [28].

The other essential issue is the quality of the available spectrum libraries, which does not fit with the epidemiology of our environment, and some authors show how they have to complement them with other home-made ones [35—38]. Theel et al. [38] describe how the error in identification of *T. rubrum* as *T. soudanense* and *T. mentagrophytes* as *T. tonsurans* are the most frequent by MS.

In summary, we can conclude here that whenever a library of reference spectra is suitable for the identification of dermatophytes, the MALDI-TOF MS identification is more economical and offers an accuracy comparable to that of DNA sequencing. The MALDI-TOF MS also represents an advantageous alternative to long and laborious dermatophytes identification in the clinical laboratory routines [37].

14.5 RAPID IDENTIFICATION IN CLINICAL SAMPLES BY MS

The possibility of detecting and identifying fungal organisms prior classical methods has been from the beginning, a very attractive target, due to the importance of establishing a directed empirical treatment [39].

The use of MALDI-TOF for the identification of fungal organisms directly from clinical samples has some disadvantages: first of all, not all samples are valid for this type of study. The ideal sample is one that comes from a usually sterile area, which could have high fungal organism concentrations, and in which there are no significant limitations in terms of sample volume, such as in blood culture and urine. In case of an infection with several fungal organisms participating, an aberrant protein spectrum can be generated, as a result of mixing several profiles or directly ignoring the organism that is in a smaller proportion.

14.5.1 Direct identification in blood cultures

Early and correct antifungal treatment in fungemia has a high impact on patients' mortality [8]. MALDI-TOF MS technique, applied directly on the pellet of a positive blood culture, provides security and speed to the identification of the species and, therefore, to the establishment of an early and adequate antifungal treatment.

Several authors show excellent results with correct identification rates ranging between 90% and 100% [40,41]. Ferroni et al. obtained above 90% of correct identifications, in candidemia with a method that selectively solubilizes the blood cells keeping the fungal membranes intact.

A fundamental aspect is the pretreatment of the sample to avoid interferences with the blood proteins before performing a protein extraction. Different protocols with good results have been described using sodium dodecyl sulfate, saponin, Tween 80, or lysis commercial systems, such as MALDI Sepsityper kit by Bruker Daltonics. This system seems to have supposed a very considerable improvement in the identification of yeasts directly from the blood culture. In a study on 42 fungemia, this processing method allowed the identification of 100% of the yeasts at the species level directly from the blood culture [42].

Some studies have suggested significant differences in the efficacy of identification by MALDI-TOF MS, among the different commercially available blood culture systems. Thus most of the studies using the BACTEC (Becton Dickinson, The United States) obtain correct identification in more than 85% of cases; some studies carried out using the BacT/Alert system (BioMérieux, France) obtain much lower results, around 30% [43−45].

14.5.2 Direct identification in other types of samples

There are few studies related to the direct identification of fungi in other types of samples. MALDI-TOF MS is a sensitive technique, but it requires high amounts of microorganisms for reliable protein profiles. Therefore, it is not expected to be a useful system, applied directly, on samples with reduced fungal loads.

Some authors have overcome this limitation by using a short incubation in liquid media, which would significantly increase their sensitivity and would continue to imply a significant saving of time in the diagnosis [46]. Although studies are lacking to systematize these protocols, it is possible that this strategy offers more attractive sensitivity, with a considerable gain in speed [7].

14.6 MS APPLICATION TO EPIDEMIOLOGICAL RESEARCH

Apart from the already contrasted capacity of MALDI-TOF MS for the identification of the fungi species, its usefulness to discriminate the different populations in epidemiological studies has been evaluated.

Protein profile generated by fungal organisms is more complex than what is needed to carry out their correct identification. Thus, there would be a series of secondary peaks, more variable within the species, that could create a *secondary profile* whose similitude were parallel to genetic proximity.

This would allow establishing levels of proximity between fungal isolates, similar to those currently established by other genetic techniques such as PFGE (pulsed-field gel electrophoresis), RAPD (random amplified polymorphic DNA), MLST (multilocus sequence typing), AFLP (amplified fragment length polymorphism), and MLP (microsatellite length polymorphism). MLP has a high discriminatory power and is frequently used in epidemiological studies and typing of different fungal species.

This approach is attractive, because if it offered similar results, it would be a significantly faster and cheaper method than the current genetic techniques [47,48].

Pulcrano et al. [49] used MALDI-TOF MS to discriminate isolates of *Candida parapsilosis* and compared their results with PFGE and MLP. MALDI-TOF had the same discriminatory power as MLP, with a good concordance between the clusters obtained by both techniques.

In a study of 104 *Candida auris* isolates from blood cultures, Prakash et al. [50] used MALDI-TOF MS to group the isolates geographically, compared to MLST and AFLP. The MS technique was able to differentiate geographically among isolates from Asia, India, and Brazil.

Dhieb et al. [51] used MALDI-TOF to classify clinical isolates of *C. glabrata* from different geographical origins compared to MLP and to see their capacity to differentiate between sensitive and resistant to fluconazole strains. The MS was able to separate the strains by their geographical origin and to differentiate sensitive and resistant strains.

De Carolis et al. [52] MALDI-TOF MS was able to discriminate species related to one another in the typing of the *C. parapsilosis* complex. However, it could not typify strains within the same species in comparison with AFLP.

MS has certain limitations. The spectra may depend on a wide variety of variables and more studies are needed to standardize MALDI-TOF system as a typing tool.

14.7 MS APPLICATION TO ANTIFUNGAL SUSCEPTIBILITY TESTING

The major antifungal drugs used in the clinic today are polyene, azole, and echinocandin classes. Drug-degrading mechanisms are not known in fungi, neither are ribosomal proteins targeted by antimycotic drugs. Most cases of decreased antifungal drug susceptibility encountered in clinical samples are rather due to species with intrinsic drug resistance, such as *C. glabrata* or *C. krusei* in the case of azoles and *C. parapsilosis* in the case of echinocandins. Also, many molds and zygomycetes show species-dependent distribution of drug susceptibility. A significant number of antifungal drug-resistant isolates are therefore predicted by MALDI-TOF simply through the identification of the fungal species.

However, under prolonged therapy or prophylaxis, isolates may additionally acquire resistance traits. In the case of azoles, these are mainly due to mutations leading to increased drug efflux or in genes involved in the ergosterol biosynthesis pathway. In the case of echinocandins, mutations occur in the gene coding for the target enzyme glucan synthase [53]. All these proteins have molecular weights that are well outside the detection range (2−20 kDa) of the current MS applications.

The detection of glucan synthase and efflux pumps is further hampered by the fact that they are multidomain membrane proteins and will most likely not be measurable under MALDI-TOF conditions. Nevertheless, optimization of ionization matrices and extraction procedures may yield approaches into this direction in the future.

A few studies use MALDI-TOF as a tool to analyze antifungal resistance. These involve growth in the presence of the drug according to the Clinical & Laboratory Standards Institute protocol and subsequent MALDI-TOF analysis of cells recovered from the wells of the microdilution plate. Differences in the mass spectra, possibly reflecting changes in the proteome or improved lysis, were seen in cells grown above the minimal inhibitory concentrations (MIC) of those particular drugs versus cells grown below.

Marinach et al. [54] observed that exposure of *Candida albicans* to different dilutions of fluconazole for 15 hours caused a change in the expression of proteins, thus varying the proteome and, therefore, the mass spectrum in those cells grown above a certain concentration of antifungal compared to those that grew below. A new cutoff point called MPCC, *minimal profile change concentration*, was thus determined. In this study, a 94% correlation was found between MIC and MPCC.

Although culture was still required and the time-saving advantage was only minimal (15 hours with MALDI-TOF versus 24 hours for a first MIC reading), this approach may serve to better discriminate isolates with trailing growth from those with true resistance.

Despite the fact that these studies show promising results, the reproducibility and methodology of the process must be optimized, standardizing the technique and defining cut points. To do this, more studies in which more strains and antifungals are evaluated are needed.

CONFLICTS OF INTERESTS

The author reports no conflicts of interests. The author alone is responsible for the content and the writing of the paper.

REFERENCES

[1] Canton R, Garcia-Rodriguez J. La espectrometria de masas MALDI-TOF en microbiologia clinica. De la innovacion a la rutina del laboratorio. Enferm Infecc Microbiol Clin 2016;34(Suppl 2):1−2.

[2] Seng P, Drancourt M, Gouriet F, et al. Ongoing revolution in bacteriology: routine identification of bacteria by matrix-assisted laser desorption ionization time-of-flight mass spectrometry. Clin Infect Dis 2009;49(4):543−51.

[3] Ayats J, Martin-Mazuelos E, Peman J, et al. Recomendaciones sobre el diagnostico de la enfermedad fungica invasora de la Sociedad Espanola de Enfermedades Infecciosas y Microbiologia Clinica (SEIMC). Actualizacion 2010. Enferm Infecc Microbiol Clin 2011;29(1) 39 e1−15.

[4] Quiles Melero I, Pelaez T, Rezusta Lopez A, Garcia-Rodriguez J. Aplicacion de la espectrometria de masas en micologia. Enferm Infecc Microbiol Clin 2016;34(Suppl 2):26−30.

[5] Clark AE, Kaleta EJ, Arora A, Wolk DM. Matrix-assisted laser desorption ionization-time of flight mass spectrometry: a fundamental shift in the routine practice of clinical microbiology. Clin Microbiol Rev 2013;26(3):547−603.

[6] Ferreira L, Vega S, Sanchez-Juanes F, et al. Identificacion bacteriana mediante espectrometria de masas matrix-assisted laser desorption ionization time-of-flight. Comparacion con la metodologia habitual en los laboratorios de Microbiologia Clinica. Enferm Infecc Microbiol Clin 2010;28(8):492−7.

[7] Muñoz Bellido JL, Gonzalez Buitrago JM. Espectrometria de masas MALDI-TOF en microbiologia clinica. Situacion actual y perspectivas futuras. Enferm Infecc Microbiol Clin 2015;33(6):369−71.

[8] Parkins MD, Sabuda DM, Elsayed S, Laupland KB. Adequacy of empirical antifungal therapy and effect on outcome among patients with invasive Candida species infections. J Antimicrob Chemother 2007;60(3):613−18.

[9] Marklein G, Josten M, Klanke U, et al. Matrix-assisted laser desorption ionization-time of flight mass spectrometry for fast and reliable identification of clinical yeast isolates. J Clin Microbiol 2009;47(9):2912−17.

[10] Wieser A, Schneider L, Jung J, Schubert S. MALDI-TOF MS in microbiological diagnostics-identification of microorganisms and beyond (mini review). Appl Microbiol Biotechnol 2012;93(3):965−74.

[11] Bader O. MALDI-TOF-MS-based species identification and typing approaches in medical mycology. Proteomics 2013;13(5):788−99.

[12] Pence MA, McElvania TeKippe E, Wallace MA, Burnham CA. Comparison and optimization of two MALDI-TOF MS platforms for the identification of medically relevant yeast species. Eur J Clin Microbiol Infect Dis 2014;33(10):1703−12.

[13] Posteraro B, De Carolis E, Vella A, Sanguinetti M. MALDI-TOF mass spectrometry in the clinical mycology laboratory: identification of fungi and beyond. Expert Rev Proteomics 2013;10(2):151−64.

[14] Vlek A, Kolecka A, Khayhan K, et al. Interlaboratory comparison of sample preparation methods, database expansions, and cutoff values for identification of yeasts by matrix-assisted laser desorption ionization-time of flight mass spectrometry using a yeast test panel. J Clin Microbiol 2014;52(8):3023−9.

[15] Siller-Ruiz M, Hernandez-Egido S, Sanchez-Juanes F, Gonzalez-Buitrago JM, Munoz-Bellido JL. Metodos rapidos de identificacion de bacterias y hongos. Espectrometria de masas MALDI-TOF, medios cromogénicos. Enferm Infecc Microbiol Clin 2017;35(5):303−13.

[16] Van Herendael BH, Bruynseels P, Bensaid M, et al. Validation of a modified algorithm for the identification of yeast isolates using matrix-assisted laser desorption/ionisation time-of-flight mass spectrometry (MALDI-TOF MS). Eur J Clin Microbiol Infect Dis 2012;31(5):841−8.

[17] Stevenson LG, Drake SK, Shea YR, Zelazny AM, Murray PR. Evaluation of matrix-assisted laser desorption ionization-time of flight mass spectrometry for identification of clinically important yeast species. J Clin Microbiol 2010;48(10):3482−6.

[18] Pinto A, Halliday C, Zahra M, et al. Matrix-assisted laser desorption ionization-time of flight mass spectrometry identification of yeasts is contingent on robust reference spectra. PLoS One 2011;6(10):e25712.

[19] Bader O, Weig M, Taverne-Ghadwal L, et al. Improved clinical laboratory identification of human pathogenic yeasts by matrix-assisted laser desorption ionization time-of-flight mass spectrometry. Clin Microbiol Infect 2011;17(9):1359−65.

[20] Bizzini A, Durussel C, Bille J, Greub G, Prod'hom G. Performance of matrix-assisted laser desorption ionization-time of flight mass spectrometry for identification of bacterial strains routinely isolated in a clinical microbiology laboratory. J Clin Microbiol 2010;48(5):1549−54.

[21] van Veen SQ, Claas EC, Kuijper EJ. High-throughput identification of bacteria and yeast by matrix-assisted laser desorption ionization-time of flight mass spectrometry in conventional medical microbiology laboratories. J Clin Microbiol 2010;48 (3):900−7.

[22] Kolecka A, Khayhan K, Groenewald M, et al. Identification of medically relevant species of arthroconidial yeasts by use of matrix-assisted laser desorption ionization-time of flight mass spectrometry. J Clin Microbiol 2013;51(8):2491−500.

[23] Ozhak-Baysan B, Ogunc D, Dogen A, Ilkit M, de Hoog GS. MALDI-TOF MS-based identification of black yeasts of the genus Exophiala. Med Mycol 2015;53 (4):347−52.

[24] Putignani L, Del Chierico F, Onori M, et al. MALDI-TOF mass spectrometry proteomic phenotyping of clinically relevant fungi. Mol Biosyst 2011;7(3):620−9.

[25] Sanguinetti M, Posteraro B. Identification of molds by matrix-assisted laser desorption ionization-time of flight mass spectrometry. J Clin Microbiol 2017;55 (2):369−79.

[26] Iriart X, Lavergne RA, Fillaux J, et al. Routine identification of medical fungi by the new Vitek MS matrix-assisted laser desorption ionization-time of flight system with a new time-effective strategy. J Clin Microbiol 2012;50(6):2107−10.

[27] Bruker Daltonics. Fungi Library—MALDI Biotyper.

[28] Cassagne C, Ranque S, Normand AC, et al. Mould routine identification in the clinical laboratory by matrix-assisted laser desorption ionization time-of-flight mass spectrometry. PLoS One 2011;6(12):e28425.

[29] Alanio A, Beretti JL, Dauphin B, et al. Matrix-assisted laser desorption ionization time-of-flight mass spectrometry for fast and accurate identification of clinically relevant Aspergillus species. Clin Microbiol Infect 2011;17(5):750−5.

[30] Lau AF, Drake SK, Calhoun LB, Henderson CM, Zelazny AM. Development of a clinically comprehensive database and a simple procedure for identification of molds from solid media by matrix-assisted laser desorption ionization-time of flight mass spectrometry. J Clin Microbiol 2013;51(3):828−34.

[31] De Carolis E, Posteraro B, Lass-Florl C, et al. Species identification of Aspergillus, Fusarium and Mucorales with direct surface analysis by matrix-assisted laser desorption ionization time-of-flight mass spectrometry. Clin Microbiol Infect 2012;18 (5):475−84.

[32] Santos C, Paterson RR, Venancio A, Lima N. Filamentous fungal characterizations by matrix-assisted laser desorption/ionization time-of-flight mass spectrometry. J Appl Microbiol 2010;108(2):375−85.

[33] Schulthess B, Ledermann R, Mouttet F, et al. Use of the Bruker MALDI Biotyper for identification of molds in the clinical mycology laboratory. J Clin Microbiol 2014;52(8):2797−803.

[34] McMullen AR, Wallace MA, Pincus DH, Wilkey K, Burnham CA. Evaluation of the Vitek MS matrix-assisted laser desorption ionization-time of flight mass spectrometry system for identification of clinically relevant filamentous fungi. J Clin Microbiol 2016;54(8):2068−73.

[35] L'Ollivier C, Cassagne C, Normand AC, et al. A MALDI-TOF MS procedure for clinical dermatophyte species identification in the routine laboratory. Med Mycol 2013;51(7):713−20.

[36] Packeu A, Hendrickx M, Beguin H, et al. Identification of the Trichophyton mentagrophytes complex species using MALDI-TOF mass spectrometry. Med Mycol 2013;51(6):580−5.

[37] L'Ollivier C, Ranque S. MALDI-TOF-based dermatophyte identification. Mycopathologia 2017;182(1−2):183−92.

[38] Theel ES, Hall L, Mandrekar J, Wengenack NL. Dermatophyte identification using matrix-assisted laser desorption ionization-time of flight mass spectrometry. J Clin Microbiol 2011;49(12):4067−71.

[39] Carlos Rodriguez J, Angel Bratos M, Merino E, Ezpeleta C. Utilizacion de MALDI-TOF en el diagnostico rapido de la sepsis. Enferm Infecc Microbiol Clin 2016;34(Suppl 2):19−25.

[40] Ferroni A, Suarez S, Beretti JL, et al. Real-time identification of bacteria and Candida species in positive blood culture broths by matrix-assisted laser desorption ionization-time of flight mass spectrometry. J Clin Microbiol 2010;48(5):1542−8.

[41] Marinach-Patrice C, Fekkar A, Atanasova R, et al. Rapid species diagnosis for invasive candidiasis using mass spectrometry. PLoS One 2010;5(1):e8862.

[42] Yan Y, He Y, Maier T, et al. Improved identification of yeast species directly from positive blood culture media by combining Sepsityper specimen processing and Microflex analysis with the matrix-assisted laser desorption ionization Biotyper system. J Clin Microbiol 2011;49(7):2528−32.

[43] Szabados F, Michels M, Kaase M, Gatermann S. The sensitivity of direct identification from positive BacT/ALERT (bioMerieux) blood culture bottles by matrix-assisted laser desorption ionization time-of-flight mass spectrometry is low. Clin Microbiol Infect 2011;17(2):192−5.

[44] Fiori B, D'Inzeo T, Di Florio V, et al. Performance of two resin-containing blood culture media in detection of bloodstream infections and in direct matrix-assisted laser desorption ionization-time of flight mass spectrometry (MALDI-TOF MS)

broth assays for isolate identification: clinical comparison of the BacT/Alert Plus and Bactec Plus systems. J Clin Microbiol 2014;52(10):3558−67.

[45] Morgenthaler NG, Kostrzewa M. Rapid identification of pathogens in positive blood culture of patients with sepsis: review and meta-analysis of the performance of the sepsityper kit. Int J Microbiol 2015;2015:827416.

[46] Kroumova V, Gobbato E, Basso E, et al. Direct identification of bacteria in blood culture by matrix-assisted laser desorption/ionization time-of-flight mass spectrometry: a new methodological approach. Rapid Commun Mass Spectrom 2011;25 (15):2247−9.

[47] Munoz Bellido JL, Vega Castano S, Ferreira L, Sanchez Juanes F, Gonzalez Buitrago JM. Aplicaciones de la proteomica en el laboratorio de Microbiologia Clinica. Enferm Infecc Microbiol Clin 2012;30(7):383−93.

[48] Sauget M, Valot B, Bertrand X, Hocquet D. Can MALDI-TOF mass spectrometry reasonably type bacteria? Trends Microbiol 2017;25(6):447−55.

[49] Pulcrano G, Roscetto E, Iula VD, et al. MALDI-TOF mass spectrometry and micro-satellite markers to evaluate Candida parapsilosis transmission in neonatal intensive care units. Eur J Clin Microbiol Infect Dis 2012;31(11):2919−28.

[50] Prakash A, Sharma C, Singh A, et al. Evidence of genotypic diversity among Candida auris isolates by multilocus sequence typing, matrix-assisted laser desorption ionization time-of-flight mass spectrometry and amplified fragment length polymorphism. Clin Microbiol Infect 2016;22(3) 277 e1−9.

[51] Dhieb C, Normand AC, Al-Yasiri M, et al. MALDI-TOF typing highlights geographical and fluconazole resistance clusters in Candida glabrata. Med Mycol 2015;53(5):462−9.

[52] De Carolis E, Hensgens LA, Vella A, et al. Identification and typing of the Candida parapsilosis complex: MALDI-TOF MS vs. AFLP. Med Mycol 2014;52(2):123−30.

[53] Pfaller MA. Antifungal drug resistance: mechanisms, epidemiology, and consequences for treatment. Am J Med 2012;125(1 Suppl):S3−13.

[54] Marinach C, Alanio A, Palous M, et al. MALDI-TOF MS-based drug susceptibility testing of pathogens: the example of *Candida albicans* and fluconazole. Proteomics 2009;9(20):4627−31.

CHAPTER FIFTEEN

Application of MALDI-TOF MS in Bacterial Strain Typing and Taxonomy

Dra Esther Culebras
Department of Clinical Microbiology, Hospital Clínico San Carlos, Madrid, Spain

15.1 INTRODUCTION

Bacterial typing involves the identification to strain level of bacterial isolates within a species. This technique seeks to establish the relationship between epidemiologically linked isolates. Microbial typing provides information not only about similarities of related isolates but also differences from those that are nonrelated, independently of their inclusion in the same microbiological species or taxon [1].

Isolate characterization is necessary to establish the extent of an outbreak, to study the sources, and to implement control procedures that can prevent the spread of resistant clones in clinical units with high-risk patients. Microbial typing has also contributed significantly to increasing the effectiveness of surveillance programs seeking to control the spread of antibiotic-resistant clones.

A good typing method uses markers that meet a number of requirements: stability during the study period, testability in every isolate, and capacity to discriminate between isolates in a reproducible way, independently of the operator, place, and time [1].

A wide variety of typing methods have been used to identify bacterial outbreaks. Current techniques can use both phenotype and genotype characteristics. Among phenotypic methods, the most common are serotyping, phagotyping, bacteriocin typing, and resitotyping. These methods differentiate between organisms in which there is a marked variation in phenotypic expression. However, most organisms causing infections make up a small subpopulation of the total strains accounting for a species and

The Use of Mass Spectrometry Technology (MALDI-TOF) in Clinical Microbiology.
DOI: https://doi.org/10.1016/B978-0-12-814451-0.00015-0

may show little diversity. Therefore, although these approaches have made valuable contributions to the confirmation and elucidation of local and national healthcare-associated outbreaks, their discrimination capacity is now considered insufficient. Also, the phenotype is not always an accurate reflection of the genotype of a microorganism; therefore, it may not be a stable epidemiological marker. In addition, all phenotypic methods show high variability, thus limiting their practical value.

Typing methods have been improved by applying molecular biology techniques to study the epidemiology of most microbial pathogens. Molecular typing techniques can be divided into two main groups, namely those based on the study of chromosomal DNA variations and those based on the amplification of specific sequences or the PCR-based method. The choice of method depends on the skill level and resources of the laboratory and the aim and scale of the investigation.

Table 15.1 shows the molecular typing methods most commonly used.

In general, although all the molecular techniques described above have proven useful, they are excessively laborious and expensive and have long lead times, thereby limiting their usefulness in real-time applications. It is in this context that MALDI-TOF MS (matrix-assisted laser desorption ionization—time-of-flight mass spectrometry) has emerged as a promising alternative for the epidemiological study of bacteria of clinical interest. This technique has the advantage of speed, low cost, and simplicity. However, to date, the implementation of this method is still underdeveloped and it involves standardized protocols and guides that help researchers to interpret the data obtained [12,13].

15.2 TYPING BY MALDI-TOF MS

Given that the ideal method for bacterial typing is fast, straightforward, and low cost, MS emerges as an attractive option to other approaches. MALDI-TOF MS is a proteotypic method whose practical aspects are simple and cause only minimal consumption cost. It has been successfully used as a rapid method for the identification of many bacterial species.

Table 15.1 Principal molecular typing methods and most relevant features of each one

Method	Description	Advantages	Drawbacks	References
RFLP	Genomic DNA digestion or of an amplicon with restriction enzymes producing short restriction fragments	All strains carrying loci homologous to probe are typeable Inexpensive, sensitive, and reproducible Good ease of interpretation	Discriminatory power depends on the choice of probes The process requires costly reagents and equipment Slow, difficult, and could take up long time to complete	[2]
AFLP	Enzyme restriction digestion of genomic DNA, binding of restriction fragments and selective amplification	Ability to use a universal protocol in combination with different restriction endonucleases Good discriminatory power, good reproducibility	Suboptimal reproducibility, particularly across different platforms Labor-intensive	[3]
MLVA	PCR amplification of loci VTR, visualizing the polymorphism to create an allele profile	May be able to differentiate suspected, fast-evolving bacterial strains from an outbreak	High assay-specificity Rapidity of repetitive DNA evolves	[4]
Rep–PCR	PCR amplification of repeated sequences in the genome	Rapid, readily available, and easy to perform Can be kit based (DiversiLab system)	Suboptimal reproducibility and discriminatory power	[5]
PFGE	Comparison of macro restriction fragments	High-discriminatory power and reproducibility Restriction profiles are easily read and interpreted	Time-consuming Requires costly reagents and equipment Labor-intensive	[6–8]
MLST	PCR amplification of housekeeping genes to create allele profiles	Typing data are readily available via the Internet and easy to compare results among laboratories and countries Good discriminatory ability	Require skilled researcher to perform. Expensive	[9]
WSG	Sequencing method to generate multiple short sequence reads across the entire genome	This technique can apply on all strains Results are reproducible with ease in interpretation High discriminatory power	The process requires costly reagents and equipment Labor-intensive	[10,11]

At present, two MALDI-TOF devices, Microflex LT (Bruker Daltonics, Bremen, Germany) and VITEK MS (bioMérieux, Marcy l'Etoile, France), are commercially available. Although both have proven useful for bacterial typing, the former requires less laboratory space, is easier to work with, and is currently the method most widely used.

The analysis workflow used for typing by MALDI-TOF MS is the same as for bacterial identification and includes culture and preparation of samples, protein extraction with ethanol—formic acid, spectrum acquisition, and data analysis. The principle of this technique is that each microorganism produces a distinct protein profile that can be used for identification purposes. This approach is attractive for typing because mass spectra commonly used for species identification can also be exploited for epidemiological typing, either directly or in an additional step of data interpretation at no additional cost.

While identification assays usually focus on peaks that are conserved across all isolates from a microbial species, high-resolution typing within a species involves the comparison of unique peaks.

Epidemiological analysis by MS can be performed in two ways, namely by using spectral libraries and/or bioinformatics approaches [14]. In library-based typing, bacteria are used to generate mass spectra, which are then compared with libraries that contain spectra of well-characterized reference bacteria. Bacterial mass spectra can be stored as reference libraries, and pattern matching or other suitable algorithms can be applied to compare mass spectral fingerprints for the identification of strains. Several differences have to be assessed in this comparison: intensity of the peak signal; presence—absence of peaks in mass spectra; and variations in the m/z value of specific peaks (peak shift) [15].

This approach is relatively simple and easy to perform. Nevertheless, variations in mass spectra may be due to growth conditions and/or sample preparation. Therefore, correct standardization of the conditions for this method is crucial.

The discrimination of strains at the below-species level by the bioinformatics approach involves identifying and distinguishing biomarkers that are exclusive to a particular strain [16]. The extraction of peaks associated with a clone from the mass spectrum enhances the performance of the typing method [15]. However, analysis by bioinformatics-enabled strategies requires the partial or complete sequencing of the microorganisms of interest. Although an increasing amount of genomic and proteomic data are available to attain strain-level characterization, there are still many microorganisms for which this information is not available. Thus, the utility of the bioinformatics approach may be limited in some settings.

There is no consensus regarding the most suitable approach for bacterial typing by MALDI-TOF MS. Although strain marker peaks have been reported [17], this information is not available for all microorganisms. Also, poor profile reproducibility can hinder the identification of strain-specific peaks, as well as MS whole-spectrum where the presence or absence of single masses, together with their intensity, is used for clustering. Therefore, it is important to standardize the methods to achieve reproducibility and mass accuracy interassays, even when these are done in different laboratories. Standardization must cover sample bacterial culturing, chemical treatment for bacterial cell wall disruption and protein extraction, and MS analysis [18].

15.2.1 Requirements for culture and sample preparation

MALDI-TOF MS is based on the analysis of phenotypic characters, and therefore distinct culture media or incubation times can cause the same species to give different mass spectra. Variation in signal intensity may reflect bacterial adaptation to culture conditions. Several studies have examined the impact of growth culture conditions and all have found variations in spectral fingerprints [19−21].

This observation highlights the relevance of well-controlled growth conditions and standardized sample preparation to obtain reproducible mass spectra. In this regard, the Clinical and Laboratory Standards Institute (CLSI) has recently published guideline methods for the identification of cultured microorganisms using MALDI-TOF MS [22]. Valentine et al. [23] and others [24] demonstrated that microorganisms cultured under different growth conditions can be correctly identified. However, to facilitate the interlaboratory comparison of results, it is important to standardize the protocol and thus avoid experimental variations. In this regard, variations in the age of the cultures, as well as those caused by the test (technical variations) and the growth conditions (biological variations), should be minimized [25]. Differences caused by any of these settings may not affect bacterial identification but could hinder bacterial typing [26].

Thus, the assessment of technical and biological reproducibility, recording of several separate spectra from each extraction, and repeating of the whole analysis (cultivation, extraction, and spectrum generation) several times are recommended. The inclusion of a sufficient number of isolates in the analysis also contributes to ensuring the validity of specific biomarker—strain associations.

A large variety of conventional solid and liquid media can be applied for typing purposes, depending on the species of interest. Liquid culture tends to provide more homogeneous bacterial populations than those grown on solid media. Also, in general, nonselective media give better results [14].

To type microorganisms properly at the subspecies level, both the culture medium used and the age of the culture are critical factors. Several studies show that the number and intensity of peaks in mass spectra change over time. These changes can reflect adaptation of bacteria to variations in culture medium with nutrient depletion and accumulation of waste products [27—29].

Without appropriate sample preparation, it is difficult to obtain spectral patterns that facilitate the identification of bacterial strains at the subspecies level. Typing through peak biomarker identification is also affected. Indeed, the growth of different bacteria can activate distinct proteins, and consequently some biomarker peaks may not be expressed under specific culture conditions. Therefore, for typing purposes, peaks expressed only in a specific medium or under certain conditions should not be considered. In addition, MS spectra include some peaks whose m/z values change in function of the degree of posttranslational modification or complexing of cofactors. These unstable peaks can be used for epidemiological purposes as long as they are well characterized [12].

Sample preparation (type of extraction, the nature of solvent, and the concentration of the matrix) also influences the mass spectrum [30]. In this regard, samples can be prepared in two ways, namely by direct bacterial transfer or by protein extraction. Bacteria can be applied directly onto a MALDI target without prior chemical treatment. However, this procedure has many drawbacks. It does not allow homogeneous distribution of the sample in the matrix and makes it difficult to use an adequate and similar amount of bacteria. The most common procedure used is the analysis of the profiles of proteins that are extracted from whole bacteria. Ethanol—formic acid extraction and trifluoroacetic acid extraction, the former being the method most frequently used, are applied to obtain proteins. After extraction, samples are embedded in a matrix, which desorbs laser energy [18]. Several matrices can be used for MALDI-TOF MS analysis, α-cyano-4-hydroxycinnamic acid (HCCA), 2,5-dihydroxybenzoic acid, and sinapinic acid (SA) being the most common. Each has its strengths and weaknesses. SA has been shown to be one of the most useful matrices [27]. However, HCCA is the most frequently used. Whatever the matrix of choice, it influences both the size and intensity of the peaks detected.

Measurement factors such as the setting of the laser shot, wear level of the laser, and quality of spots on the MALDI target plate can lead to variation in peak intensity and therefore must be considered for interlaboratory comparison purposes [15].

MALDI-TOF MS differentiation to the subspecies level can benefit both from combination with other techniques and from the use of extraction techniques or chemical modification [15]. Sample prefractionation, bacterial cell wall disruption, and protein extraction allow the enrichment of proteins and peptides [31,32]. The ability to differentiate closely related bacteria by mass patterns can also be improved by treating bacterial samples with the protease trypsin [33].

15.2.2 Data acquisition and spectrum quality

The acquisition of spectra by MALDI-TOF MS must be optimized to ensure the reliable characterization of strains. Data collection affects the quality and reproducibility of the spectra; therefore, standardization of the process is vital.

Mass spectra can be acquired either manually or automatically. Although automatic data acquisition yields less reproducible spectra than manual acquisition [34], the suitable configuration of automatic acquisition settings significantly improves fingerprint-based approaches [35].

The parameters to be optimized include peak selection mass range, signal-to-noise ratio (S/N), base peak intensity, base peak minimum resolution, and number of shots summed. In addition, the laser power and laser movement on the sample need to be controlled. After optimization, Zhang and colleagues [35] found that the reproducibility of replicate spectra exceeded those obtained manually for *Pseudomonas aeruginosa*, *Klebsiella pneumoniae*, and *Serratia marcescens*. These observations thus indicate that similar results can be obtained with other bacterial species.

Regardless of the mass spectrometer and the method used, spectra are typically collected in the linear positive mode at a laser frequency of 20 Hz in the range of 2–20 kDa, although broader and narrower ranges have been used with success [36–38].

Generally, spectra are considered high quality when peaks have a significant intensity and S/N ratio. To reduce the technical variability among spectra, several replicates are required. Each replicate is constructed by accumulating enough laser shots to guarantee test reproducibility. Spectra are externally calibrated using Bacterial Test Standard and preprocessed by smoothing, baseline subtraction, and internal peak alignment. Internal recalibration is essential for typing because it allows the reliable comparison of spectra.

Grouping spectra from multiple measurements allows the calculation of the so-called main spectra (MSP, BioTyper) or SuperSpectrum (SARAMIS). Creation of MSP using the appropriate parameters circumvents biological variability and facilitates dendrogram construction, thus enhancing the performance of epidemiological studies. To ensure good quality MSP and dendrograms, flat liners and poor-quality spectra should be excluded from further processing.

15.2.3 Data analysis

Bacterial profiling at the strain level involves the generation of complex data sets, thus underscoring the need for software-assisted data analysis.

Each of the two main MALDI-TOF MS instruments has its own software solution. Bruker uses MALDI BioTyper while Shimadzu uses the Shimadzu Launchpad software with the SARAMIS database. The commercial software tools are usually integrated with their own spectra reference database, and they apply a unique algorithm for spectra processing, pattern matching, and result interpretation. A search of PubMed shows that Bruker software is the most widely used in typing studies.

Once the raw spectra of an isolate have been preprocessed in Biotyper Software and the MPS have been created, several algorithms can be used to achieve the taxonomic classification of bacterial strains. Of note, peak data obtained by MALDI-TOF MS can be processed using various software packages and algorithms. Consequently, there is great diversity in the classification of bacterial strains, thereby hindering the comparison of results between studies.

Basic BioTyper Bruker principal component analysis (PCA) and composite correlation index (CCI) are the first easy-to-use starting parameters for clustering purposes.

The CCI is a statistical method for analyzing the relationships between spectra. In this analysis, raw spectra are divided into several intervals of the same size, and the composition of cross-correlations and autocorrelations of all intervals (in terms of geometric mean) is used as a distance parameter between the spectra. CCI values of around 1 indicate high conformance of spectra while values near 0 indicate diversity. The results are translated into a heat map where closely related spectra are marked in hot colors and unrelated ones in cold colors [39].

PCA is a widely used mathematical technique designed to extract, display, and rank variance within a data set. PCA hierarchical clustering determines clusters with similar protein expression by applying Euclidean distance measures and single linkage algorithms. PCA takes into consideration peak mass and peak intensities and therefore is highly sensitive to culture conditions [40].

Other methods can be used to visualize epidemiological relations. UPGMA (Unweighted Pair Group Method with Arithmetic Mean) is a straightforward approach that employs a sequential clustering algorithm to construct a phylogenetic tree from distance matrix taxa [41,42]. Similarities between two lists of peaks can be represented as a dendrogram by applying the single link-age cluster algorithm [43].

Many programs offer other methods of nonhierarchical grouping, such as k-means and self-organized maps. These types of algorithm are more reliable and robust but they are not capable of generating dendrograms.

One of the most interesting bioinformatics tool for bacterial strain typing and taxonomical classification is the ClinProTools software. This statistical program includes a number of highly sophisticated mathematical algorithms and is specifically designed to process peak data obtained by MALDI-TOF MS and generate pattern recognition models in a relatively rapid and flexible manner [44]. Nevertheless, although ClinProTools provides satisfactory results [45,46], its software is not included in the standard package and is therefore not freely available to some laboratories. This explains why researchers use a variety of mathematical algorithms that are freely available for MS data analysis purposes.

Bacterial typing on the basis of a sequence of peaks can be modeled as a classification problem. In this context, "classification" means identifying the set of categories to which a new observation belongs, given a labeled data set whose categories are known [47]. Mathematical and bioinformatics approaches can potentially provide a reproducible and classify bacteria [46,48]. In this regard, Support Vector Machine and neural networks have been used to identify bacterial clones [49–51]. Other algorithms used for classification purposes include the K-Nearest-Neighbors [52], the Decision Tree [53], and Random Forests [54].

15.3 OVERVIEW OF PUBLISHED STUDIES

15.3.1 Gram-negative bacteria

Gram-negative microorganisms are a serious public health problem, causing infections of medical significance, including a large proportion of nosocomial infections. In this regard, the *Enterobacteriaceae* family is the most commonly identified group. However, other multidrug-resistant microorganisms, including *P. aeruginosa* and *Acinetobacter baumannii*, are increasingly reported [55–57].

Given that most Gram-negative bacilli isolated in hospital-acquired infections are resistant to many antibiotics, rapid and reliable identification and typing are crucial.

15.3.1.1 Enterobacteriaceae

Many applications have been developed for *Enterobacteriaceae*. These combine pure epidemiology with the characterization of resistance proteins, and they can facilitate the tracking of plasmids more than strains [58,59].

One of the most widely studied species is probably *Escherichia coli*, as reflected by the number of studies addressing this strain [58,60–64]. *Escherichia coli* is a ubiquitous human pathogen that has registered outbreaks worldwide. Most MS typing studies of *E. coli* have focused on the detection of the isolates involved in outbreaks. The identification of characteristic marker peaks emerges as a suitable method to achieve this objective, and satisfactory results have been achieved using this approach [17,37].

However, although the detection of isolates involved in outbreaks is of great importance for the control of infection, from the epidemiological point of view, it is also of interest to distinguish clonal complexes. Various authors have successfully used MS fingerprinting analysis for identifying and clustering *E. coli* clones [64,65].

These data indicate that MALDI-TOF MS is a reliable method through which to monitor the epidemiology of *E. coli* clones of clinical relevance.

The differentiation between *E. coli* and *Shigella* species has been addressed using MALDI-TOF MS and ClinProTools software. The combination of these tools allowed the identification of biomarker peaks and the generation of models to differentiate the two microorganisms [66,67].

Another member of *Enterobacteriaceae* responsible for nosocomial outbreaks worldwide is *Klebsiella pneumoniae* [51]. The rapid identification and detection of *K. pneumoniae*, mainly carbapenemase-producing, is crucial to ascertain outbreaks, as well as to limit its spread. However, the results of MS typing for *K. pneumoniae* epidemiological analysis are contradictory [68–71].

Studies with *Salmonella* [72–74], *Enterobacter* [75], and *Serratia* [76] also highlight the utility of MALDI-TOF for *Enterobacteriaceae* typing.

15.3.1.2 Nonfermenting Gram-negative bacteria

Nonfermenting Gram-negative bacteria have emerged as important healthcare-associated pathogens as they exhibit intrinsic multidrug resistance. These organisms have the potential to spread horizontally and cause opportunistic infections in the clinical setting, mainly in immunocompromised patients.

MS typing of nonfermenting Gram-negative bacteria is inconclusive. Cabrolier et al. [77] identified the five high-risk clones of *P. aeruginosa* (sequence type 111 (ST111), ST175, ST235, ST253, and ST395) accurately and quickly, while other authors [78] obtained unassured differentiation between biotyper groups in this bacterium. The use of more sophisticated MS processes improves the resolution [79,80].

Similar results were obtained with *Acinetobacter* species [81,82]. It is possible to find specific peaks to clearly identify new species in the *A. baumannii* group [57,83] and to cluster isolates of diverse origins [84]. Also, MS has been used to detect outbreaks associated with *A. baumannii* isolates [81]. Nevertheless, other studies found insufficient discriminatory power of MALDI-TOF MS to determine *A. baumannii* clonality [85,86].

The advantage of MALDI-TOF MS typing over traditional typing techniques in the analysis of other nonfermenting Gram-negative bacteria has not been established. Several studies have evaluated the techniques with *Burkholderia cepacia* [87], *Burkholderia pseudomallei* [88], *Achromobacter* [89], *Stenotrophomonas maltophilia* [90].

15.3.2 Gram-positive bacteria

15.3.2.1 Staphylococcus aureus

Methicillin-resistant *Staphylococcus aureus* (MRSA) is an important pathogen associated with nosocomial infections. In this regard, isolate typing is required to analyze outbreaks and establish adequate surveillance programs. MS has provided acceptable results in this context [91,92]. Several studies have reported on the capacity of MALDI-TOF MS to correctly distinguish the main *S. aureus* clonal complexes (CC5, CC8, CC22, CC30, and CC45) [49,50,93,94]. Knowledge of the genetics of MRSA allows the correlation of peak shifts in the main clonal lineage with point mutations in the respective genes [95]. A protocol has been successfully applied by local epidemiology and outbreak control [96]. However, the main issue with MRSA typing is the heterogeneity in the results of different laboratories. Despite this, the results with MRSA are promising.

15.3.2.2 *Enterococcus spp.*

Enterococci have long been recognized as important human pathogens, especially in the nosocomial environment. *Enterococcus faecalis* and *Enterococcus faecium* are the most prevalent species found in clinical settings. Stains of the latter are usually resistant to multiple antibiotics; therefore, early detection and typing of this microorganism is vital. In this regard, MALDI-TOF has provided contradictory results for enterococci. Nakano et al. [97] obtained a high detection rate of vanA-positive isolates, and Grifflin et al. [98] reported similar results with vanB producers. However, others were not able either to identify relevant marker peaks linked to glycopeptide-resistance determinants (vanA, vanB) [92] or to reliably differentiate *E. faecium* clones or clonal complexes. The identification of high-risk *E. faecium* clones has been successfully achieved recently [99]; however, whole genome sequencing outperforms MALDI-TOF with respect to definitive determination of the evolutionary distance between enterococci isolates [100].

15.3.3 Other microorganisms

Most relevant studies are focused on *Streptococcus pneumoniae* serotypes. The determination of capsular types is key to preventing the spread of invasive pneumococcal diseases and to estimating the impact of pneumococcal conjugate vaccines. After optimizing the settings, MALDI-TOF MS shows greater potential for serotyping pneumococcal strains and in a more straightforward manner than traditional methods [101,102].

Other Gram-positive microorganisms have been studied for taxonomic purposes using MS. In this regard, this technique has produced variable results for *Streptococcus agalactiae* [103] *Streptococcus pyogenes* [104], *Staphylococcus saprophyticus* [105], *Bacillus* spp. [106,107], *Listeria* [108], and others [109,110].

MALDI-TOF MS—based typing has also been used for the epidemiological analysis of *Helicobacter cinaedi* [111], *Clostridium difficile* [112], *Campylobacter jejuni* [40], *Mycobacterium abscessus* [113,114], *Corynebacterium striatum* [115], *Bacteroides fragilis* [116], *Yersinia enterocolitica* [117], *Candida auris* [118], and several other microorganisms [119,120] with variable results.

15.4 REMARKS

There is no commercial system currently available for bacterial typing. However, the number of typing studies performed by MS is increasing; they do not always give satisfactory results. While MS typing of *Enterobacteriaceae* and *S. aureus* provides attractive results, those of other microorganisms are commonly not as conclusive. An improvement in the results is often directly related to the development of bioinformatics tools and algorithm models that allow the analysis of large amounts of data. To date, subspecies-level discrimination has been pursued mainly using in-house algorithms and workflows. Therefore, interlaboratory comparison of results is often hindered. The development of standardized methods, the maintenance of reliable reference databases, and the identification of discriminatory mass signals will allow for homogeneous interlaboratory results. In this future context, the use of MALDI-TOF MS for bacterial typing and taxonomy is expected to gain ground in clinical microbiology.

REFERENCES

[1] van Belkum A, Tassios PT, Dijkshoorn L, Haeggman S, Cookson B, Fry NK, et al. Guidelines for the validation and application of typing methods for use in bacterial epidemiology. Clin Microbiol Infect 2007;13(Suppl 3):1–46.

[2] Dai S, Long Y. Genotyping analysis using an RFLP assay. Methods Mol Biol 2015;1245:91–9.

[3] Janssen P, Coopman R, Huys G, Swings J, Bleeker M, Vos P, et al. Evaluation of the DNA fingerprinting method AFLP as an new tool in bacterial taxonomy. Microbiology 1996;142(Pt 7):1881–93.

[4] Nadon CA, Trees E, Ng LK, Moller Nielsen E, Reimer A, Maxwell N, et al. Development and application of MLVA methods as a tool for inter-laboratory surveillance. Euro Surveill 2013;18:20565.

[5] Nowak J, Zander E, Stefanik D, Higgins PG, Roca I, Vila J, et al. High incidence of pandrug-resistant *Acinetobacter baumannii* isolates collected from patients with ventilator-associated pneumonia in Greece, Italy and Spain as part of the MagicBullet clinical trial. J Antimicrob Chemother 2017;72:3277–82.

[6] Golle A, Janezic S, Rupnik M. Low overlap between carbapenem resistant *Pseudomonas aeruginosa* genotypes isolated from hospitalized patients and wastewater treatment plants. PLoS One 2017;12:e0186736.

[7] Costa CL, Mano de Carvalho CB, Gonzalez RH, Gifoni MAC, Ribeiro RA, Quesada-Gomez C, et al. Molecular epidemiology of *Clostridium difficile* infection in a Brazilian cancer hospital. Anaerobe 2017;48:232−6.

[8] Solgi H, Giske CG, Badmasti F, Aghamohammad S, Havaei SA, Sabeti S, et al. Emergence of carbapenem resistant *Escherichia coli* isolates producing blaNDM and blaOXA-48-like carried on IncA/C and IncL/M plasmids at two Iranian university hospitals. Infect Genet Evol 2017;55:318−23.

[9] Maiden MC, Bygraves JA, Feil E, Morelli G, Russell JE, Urwin R, et al. Multilocus sequence typing: a portable approach to the identification of clones within populations of pathogenic microorganisms. Proc Natl Acad Sci U S A 1998;95:3140−5.

[10] Turner CE, Bedford L, Brown NM, Judge K, Torok ME, Parkhill J, et al. Community outbreaks of group A Streptococcus revealed by genome sequencing. Sci Rep 2017;7:8554.

[11] Mosites E, Frick A, Gounder P, Castrodale L, Li Y, Rudolph K, et al. Outbreak of invasive infections from subtype emm26.3 group A Streptococcus among homeless adults-Anchorage, Alaska, 2016-2017. Clin Infect Dis 2018;66(7):1068−74. Available from: https://doi.org/10.1093/cid/cix921.

[12] Spinali S, van Belkum A, Goering RV, Girard V, Welker M, Van Nuenen M, et al. Microbial typing by matrix-assisted laser desorption ionization-time of flight mass spectrometry: do we need guidance for data interpretation? J Clin Microbiol 2015;53:760−5.

[13] Murray PR. Matrix-assisted laser desorption ionization time-of-flight mass spectrometry; usefulness for taxonomy and epidemiology. Clin Microbiol Infect 2010;16:1626−30.

[14] Sandrin TR, Goldstein JE, Schumaker S. MALDI TOF MS profiling of bacteria at the strain level: a review. Mass Spectrom Rev 2013;32:188−217.

[15] Sauget M, Valot B, Bertrand X, Hocquet D. Can MALDI-TOF mass spectrometry reasonably type bacteria? Trends Microbiol 2017;25:447−55.

[16] Zautner AE, Masanta WO, Weig M, Gross U, Bader O. Mass spectrometry-based phylo proteomics (MSPP): a novel microbial typing method. Sci Rep 2015;5:13431.

[17] Christner M, Trusch M, Rohde H, Kwiatkowski M, Schluter H, Wolters M, et al. Rapid MALDI-TOF mass spectrometry strain typing during a large outbreak of Shiga-toxigenic *Escherichia coli*. PLoS One 2014;9:e101924.

[18] Freiwald A, Sauer S. Phylogenetic classification and identification of bacteria by mass spectrometry. Nat Protoc 2009;4:732−42.

[19] Arnold RJ, Karty JA, Ellington AD, Reilly JP. Monitoring the growth of a bacteria culture by MALDI-MS of whole cells. Anal Chem 1999;71:1990−6.

[20] Longo MA, Novella IS, Garcia LA, Diaz M. Comparison of Bacillus subtilis and Serratia marcescens as protease producers under different operating conditions. J Biosci Bioeng 1999;88:35−40.

[21] Carbonnelle E, Mesquita C, Bille E, Day N, Dauphin B, Beretti JL, et al. MALDI-TOF mass spectrometry tools for bacterial identification in clinical microbiology laboratory. Clin Biochem 2011;44:104−9.

[22] CLSI. Methods for the identification of cultured microorganisms using matrix-assisted laser desorption/ionization time-of-flight mass spectrometry. 1st ed. Clinical and Laboratory Standards Institute Guideline; 2017. M58.

[23] Valentine N, Wunschel S, Wunschel D, Petersen C, Wahl K. Effect of culture conditions on microorganism identification by matrix-assisted laser desorption ionization mass spectrometry. Appl Environ Microbiol 2005;71:58−64.

[24] Bernardo K, Pakulat N, Macht M, Krut O, Seifert H, Fleer S, et al. Identification and discrimination of Staphylococcus aureus strains using matrix-assisted laser desorption/ionization-time of flight mass spectrometry. Proteomics 2002;2:747—53.

[25] Sloan A, Wang G, Cheng K. Traditional approaches versus mass spectrometry in bacterial identification and typing. Clin Chim Acta 2017;473:180—5.

[26] Sauer S, Freiwald A, Maier T, Kube M, Reinhardt R, Kostrzewa M, et al. Classification and identification of bacteria by mass spectrometry and computational analysis. PLoS One 2008;3:e2843.

[27] Giebel R, Worden C, Rust SM, Kleinheinz GT, Robbins M, Sandrin TR. Microbial fingerprinting using matrix-assisted laser desorption ionization time-of-flight mass spectrometry (MALDI-TOF MS) applications and challenges. Adv Appl Microbiol 2010;71:149—84.

[28] Vargha M, Takats Z, Konopka A, Nakatsu CH. Optimization of MALDI-TOF MS for strain level differentiation of Arthrobacter isolates. J Microbiol Methods 2006;66:399—409.

[29] Ruelle V, El Moualij B, Zorzi W, Ledent P, Pauw ED. Rapid identification of environmental bacterial strains by matrix-assisted laser desorption/ionization time-of-flight mass spectrometry. Rapid Commun Mass Spectrom 2004;18:2013—19.

[30] Sauer S, Kliem M. Mass spectrometry tools for the classification and identification of bacteria. Nat Rev Microbiol 2010;8:74—82.

[31] Fenselau C, Demirev PA. Characterization of intact microorganisms by MALDI mass spectrometry. Mass Spectrom Rev 2001;20:157—71.

[32] Lay Jr. JO. MALDI-TOF mass spectrometry of bacteria. Mass Spectrom Rev 2001;20:172—94.

[33] Schmidt F, Fiege T, Hustoft HK, Kneist S, Thiede B. Shotgun mass mapping of Lactobacillus species and subspecies from caries related isolates by MALDI-MS. Proteomics 2009;9:1994—2003.

[34] Schumaker S, Borror CM, Sandrin TR. Automating data acquisition affects mass spectrum quality and reproducibility during bacterial profiling using an intact cell sample preparation method with matrix-assisted laser desorption/ionization time-of-flight mass spectrometry. Rapid Commun Mass Spectrom 2012;26:243—53.

[35] Zhang L, Borror CM, Sandrin TR. A designed experiments approach to optimization of automated data acquisition during characterization of bacteria with MALDI-TOF mass spectrometry. PLoS One 2014;9:e92720.

[36] Giebel RA, Fredenberg W, Sandrin TR. Characterization of environmental isolates of Enterococcus spp. by matrix-assisted laser desorption/ionization time-of-flight mass spectrometry. Water Res 2008;42:931—40.

[37] Christner M, Dressler D, Andrian M, Reule C, Petrini O. Identification of Shiga-toxigenic Escherichia coli outbreak isolates by a novel data analysis tool after matrix-assisted laser desorption/ionization time-of-flight mass spectrometry. PLoS One 2017;12:e0182962.

[38] Dickinson DN, La Duc MT, Haskins WE, Gornushkin I, Winefordner JD, Powell DH, et al. Species differentiation of a diverse suite of Bacillus spores by mass spectrometry-based protein profiling. Appl Environ Microbiol 2004;70:475—82.

[39] Mayer-Scholl A, Murugaiyan J, Neumann J, Bahn P, Reckinger S, Nockler K. Rapid identification of the foodborne pathogen Trichinella spp. by matrix-assisted laser desorption/ionization mass spectrometry. PLoS One 2016;11:e0152062.

[40] Zautner AE, Masanta WO, Tareen AM, Weig M, Lugert R, Gross U, et al. Discrimination of multilocus sequence typing-based Campylobacter jejuni subgroups by MALDI-TOF mass spectrometry. BMC Microbiol 2013;13:247.

[41] Teramoto K, Kitagawa W, Sato H, Torimura M, Tamura T, Tao H. Phylogenetic analysis of Rhodococcus erythropolis based on the variation of ribosomal proteins as

observed by matrix-assisted laser desorption ionization-mass spectrometry without using genome information. J Biosci Bioeng 2009;108:348−53.

[42] Teramoto K, Sato H, Sun L, Torimura M, Tao H, Yoshikawa H, et al. Phylogenetic classification of Pseudomonas putida strains by MALDI-MS using ribosomal subunit proteins as biomarkers. Anal Chem 2007;79:8712−19.

[43] Conway GC, Smole SC, Sarracino DA, Arbeit RD, Leopold PE. Phyloproteomics: species identification of Enterobacteriaceae using matrix-assisted laser desorption/ionization time-of-flight mass spectrometry. J Mol Microbiol Biotechnol 2001;3:103−12.

[44] Ikryannikova LN, Filimonova AV, Malakhova MV, Savinova T, Filimonova O, Ilina EN, et al. Discrimination between Streptococcus pneumoniae and Streptococcus mitis based on sorting of their MALDI mass spectra. Clin Microbiol Infect 2013;19:1066−71.

[45] Kornienko M, Ilina E, Lubasovskaya L, Priputnevich T, Falova O, Sukhikh G, et al. Analysis of nosocomial Staphylococcus haemolyticus by MLST and MALDI-TOF mass spectrometry. Infect Genet Evol 2016;39:99−105.

[46] Oberle M, Wohlwend N, Jonas D, Maurer FP, Jost G, Tschudin-Sutter S, et al. The technical and biological reproducibility of matrix-assisted laser desorption ionization-time of flight mass spectrometry (MALDI-TOF MS) based typing: employment of bioinformatics in a multicenter study. PLoS One 2016;11:e0164260.

[47] Bădică C, Nguyễn TN, Brezovan M. 2013. Computational collective intelligence. Technologies and applications. In: 5th International Conference. Craiova, Romania: ICCCI; September 11−13, 2013.

[48] De Bruyne K, Slabbinck B, Waegeman W, Vauterin P, De Baets B, Vandamme P. Bacterial species identification from MALDI-TOF mass spectra through data analysis and machine learning. Syst Appl Microbiol 2011;34:20−9.

[49] Sauget M, van der Mee-Marquet N, Bertrand X, Hocquet D. Matrix-assisted laser desorption ionization-time of flight mass spectrometry can detect Staphylococcus aureus clonal complex 398. J Microbiol Methods 2016;127:20−3.

[50] Camoez M, Sierra JM, Dominguez MA, Ferrer-Navarro M, Vila J, Roca I. Automated categorization of methicillin-resistant Staphylococcus aureus clinical isolates into different clonal complexes by MALDI-TOF mass spectrometry. Clin Microbiol Infect 2016;22 161.e1−161.e7.

[51] Angeletti S, Dicuonzo G, Lo Presti A, Cella E, Crea F, Avola A, et al. MALDI-TOF mass spectrometry and blakpc gene phylogenetic analysis of an outbreak of carbapenem-resistant K. pneumoniae strains. New Microbiol 2015;38:541−50.

[52] Park HS, Rinehart MT, Walzer KA, Chi JT, Wax A. Automated detection of P. falciparum using machine learning algorithms with quantitative phase images of unstained cells. PLoS One 2016;11 e0163045.

[53] Sui M, Huang X, Li Y, Ma X, Zhang C, Li X, et al. Application and comparison of laboratory parameters for forecasting severe hand-foot-mouth disease using logistic regression, discriminant analysis and decision tree. Clin Lab 2016;62:1023−31.

[54] Chen J, Wright K, Davis JM, Jeraldo P, Marietta EV, Murray J, et al. An expansion of rare lineage intestinal microbes characterizes rheumatoid arthritis. Genome Med 2016;8:43.

[55] Hackel MA, Tsuji M, Yamano Y, Echols R, Karlowsky JA, Sahm DF. In vitro activity of the siderophore cephalosporin, cefiderocol, against carbapenem-non-susceptible and multidrug-resistant isolates of gram-negative bacilli collected worldwide in 2014-2016. Antimicrob Agents Chemother 2018;62(2). Available from: https://doi.org/10.1128/AAC.01968-17.

[56] Kengkla K, Kongpakwattana K, Saokaew S, Apisarnthanarak A, Chaiyakunapruk N. Comparative efficacy and safety of treatment options for MDR and XDR

Acinetobacter baumannii infections: a systematic review and network meta-analysis. J Antimicrob Chemother 2018;73(1):22−32. Available from: https://doi.org/10.1093/jac/dkx368.

[57] Espinal P, Seifert H, Dijkshoorn L, Vila J, Roca I. Rapid and accurate identification of genomic species from the Acinetobacter baumannii (Ab) group by MALDI-TOF MS. Clin Microbiol Infect 2012;18:1097−103.

[58] Egli A, Tschudin-Sutter S, Oberle M, Goldenberger D, Frei R, Widmer AF. Matrix-assisted laser desorption/ionization time of flight mass-spectrometry (MALDI-TOF MS) based typing of extended-spectrum beta-lactamase producing E. coli—a novel tool for real-time outbreak investigation. PLoS One 2015;10: e0120624.

[59] Lau AF, Wang H, Weingarten RA, Drake SK, Suffredini AF, Garfield MK, et al. A rapid matrix-assisted laser desorption ionization-time of flight mass spectrometry-based method for single-plasmid tracking in an outbreak of carbapenem-resistant Enterobacteriaceae. J Clin Microbiol 2014;52:2804−12.

[60] Fagerquist CK, Zaragoza WJ, Sultan O, Woo N, Quinones B, Cooley MB, et al. Top-down proteomic identification of Shiga toxin 2 subtypes from Shiga toxin-producing Escherichia coli by matrix-assisted laser desorption ionization-tandem time of flight mass spectrometry. Appl Environ Microbiol 2014;80:2928−40.

[61] Nakamura A, Komatsu M, Kondo A, Ohno Y, Kohno H, Nakamura F, et al. Rapid detection of B2-ST131 clonal group of extended-spectrum beta-lactamase-producing Escherichia coli by matrix-assisted laser desorption ionization-time-of-flight mass spectrometry: discovery of a peculiar amino acid substitution in B2-ST131 clonal group. Diagn Microbiol Infect Dis 2015;83:237−44.

[62] Matsumura Y, Yamamoto M, Nagao M, Tanaka M, Takakura S, Ichiyama S. Detection of Escherichia coli sequence type 131 clonal group among extended-spectrum beta-lactamase-producing E. coli using VITEK MS Plus matrix-assisted laser desorption ionization-time of flight mass spectrometry. J Microbiol Methods 2015;119:7−9.

[63] Lafolie J, Sauget M, Cabrolier N, Hocquet D, Bertrand X. Detection of Escherichia coli sequence type 131 by matrix-assisted laser desorption ionization time-of-flight mass spectrometry: implications for infection control policies? J Hosp Infect 2015;90:208−12.

[64] Matsumura Y, Yamamoto M, Nagao M, Tanaka M, Machida K, Ito Y, et al. Detection of extended-spectrum-beta-lactamase-producing Escherichia coli ST131 and ST405 clonal groups by matrix-assisted laser desorption ionization-time-of flight mass spectrometry. J Clin Microbiol 2014;52:1034−40.

[65] Novais A, Sousa C, de Dios Caballero J, Fernandez-Olmos A, Lopes J, Ramos H, et al. MALDI-TOF mass spectrometry as a tool for the discrimination of high-risk Escherichia coli clones from phylogenetic groups B2 (ST131) and D (ST69, ST405, ST393). Eur J Clin Microbiol Infect Dis 2014;33:1391−9.

[66] Khot PD, Fisher MA. Novel approach for differentiating Shigella species and Escherichia coli by matrix-assisted laser desorption ionization-time of flight mass spectrometry. J Clin Microbiol 2013;51:3711−16.

[67] Paauw A, Jonker D, Roeselers G, Heng JM, Mars-Groenendijk RH, Trip H, et al. Rapid and reliable discrimination between Shigella species and Escherichia coli using MALDI-TOF mass spectrometry. Int J Med Microbiol 2015;305:446−52.

[68] Sakarikou C, Ciotti M, Dolfa C, Angeletti S, Favalli C. Rapid detection of carbapenemase-producing Klebsiella pneumoniae strains derived from blood cultures by matrix-assisted laser desorption ionization-time of flight mass spectrometry (MALDI-TOF MS). BMC Microbiol 2017;17:54.

[69] Rodrigues C, Novais A, Sousa C, Ramos H, Coque TM, Canton R, et al. Elucidating constraints for differentiation of major human Klebsiella pneumoniae clones using MALDI-TOF MS. Eur J Clin Microbiol Infect Dis 2017;36:379—86.

[70] Berrazeg M, Diene SM, Drissi M, Kempf M, Richet H, Landraud L, et al. Biotyping of multidrug-resistant Klebsiella pneumoniae clinical isolates from France and Algeria using MALDI-TOF MS. PLoS One 2013;8:e61428.

[71] Sachse S, Bresan S, Erhard M, Edel B, Pfister W, Saupe A, et al. Comparison of multilocus sequence typing, RAPD, and MALDI-TOF mass spectrometry for typing of beta-lactam-resistant Klebsiella pneumoniae strains. Diagn Microbiol Infect Dis 2014;80:267—71.

[72] Dieckmann R, Malorny B. Rapid screening of epidemiologically important Salmonella enterica subsp. enterica serovars by whole-cell matrix-assisted laser desorption ionization-time of flight mass spectrometry. Appl Environ Microbiol 2011;77:4136—46.

[73] Kuhns M, Zautner AE, Rabsch W, Zimmermann O, Weig M, Bader O, et al. Rapid discrimination of Salmonella enterica serovar Typhi from other serovars by MALDI-TOF mass spectrometry. PLoS One 2012;7:e40004.

[74] Ojima-Kato T, Yamamoto N, Nagai S, Shima K, Akiyama Y, Ota J, et al. Application of proteotyping Strain Solution ver. 2 software and theoretically calculated mass database in MALDI-TOF MS typing of Salmonella serotype. Appl Microbiol Biotechnol 2017;101:8557—69.

[75] Khennouchi NC, Loucif L, Boutefnouchet N, Allag H, Rolain JM. MALDI-TOF MS as a tool to detect a nosocomial outbreak of extended-spectrum-beta-lactamase- and ArmA methyltransferase-producing Enterobacter cloacae clinical isolates in Algeria. Antimicrob Agents Chemother 2015;59:6477—83.

[76] Batah R, Loucif L, Olaitan AO, Boutefnouchet N, Allag H, Rolain JM. Outbreak of Serratia marcescens coproducing ArmA and CTX-M-15 mediated high levels of resistance to aminoglycoside and extended-spectrum beta-lactamases, Algeria. Microb Drug Resist 2015;21:470—6.

[77] Cabrolier N, Sauget M, Bertrand X, Hocquet D. Matrix-assisted laser desorption ionization-time of flight mass spectrometry identifies Pseudomonas aeruginosa high-risk clones. J Clin Microbiol 2015;53:1395—8.

[78] Oumerci T, Jensen V, Talbot SR, Hofmann W, Kostrzewa M, Schlegelberger B, et al. Comprehensive MALDI-TOF biotyping of the non-redundant Harvard Pseudomonas aeruginosa PA14 transposon insertion mutant library. PLoS One 2015;10:e0117144.

[79] Fleurbaaij F, Kraakman ME, Claas EC, Knetsch CW, van Leeuwen HC, van der Burgt YE, et al. Typing Pseudomonas aeruginosa isolates with ultrahigh resolution MALDI-FTICR mass spectrometry. Anal Chem 2016;88:5996—6003.

[80] Mulet M, Gomila M, Ramirez A, Cardew S, Moore ER, Lalucat J, et al. Uncommonly isolated clinical Pseudomonas: identification and phylogenetic assignation. Eur J Clin Microbiol Infect Dis 2017;36:351—9.

[81] Mencacci A, Monari C, Leli C, Merlini L, De Carolis E, Vella A, et al. Typing of nosocomial outbreaks of Acinetobacter baumannii by use of matrix-assisted laser desorption ionization-time of flight mass spectrometry. J Clin Microbiol 2013;51:603—6.

[82] Wang H, Drake SK, Yong C, Gucek M, Tropea M, Rosenberg AZ, et al. A novel peptidomic approach to strain typing of clinical Acinetobacter baumannii isolates using mass spectrometry. Clin Chem 2016;62:866—75.

[83] Mari-Almirall M, Cosgaya C, Higgins PG, Van Assche A, Telli M, Huys G, et al. MALDI-TOF/MS identification of species from the Acinetobacter baumannii (Ab)

group revisited: inclusion of the novel A. seifertii and A. dijkshoorniae species. Clin Microbiol Infect 2017;23 210 e211-210 e219.

[84] Elbehiry A, Marzouk E, Hamada M, Al-Dubaib M, Alyamani E, Moussa IM. Application of MALDI-TOF MS fingerprinting as a quick tool for identification and clustering of foodborne pathogens isolated from food products. New Microbiol 2017;40:269−78.

[85] Sousa C, Botelho J, Grosso F, Silva L, Lopes J, Peixe L. Unsuitability of MALDI-TOF MS to discriminate Acinetobacter baumannii clones under routine experimental conditions. Front Microbiol 2015;6:481.

[86] Rim JH, Lee Y, Hong SK, Park Y, Kim M, D'Souza R, et al. Insufficient discriminatory power of matrix-assisted laser desorption ionization time-of-flight mass spectrometry dendrograms to determine the clonality of multi-drug-resistant Acinetobacter baumannii isolates from an intensive care unit. Biomed Res Int 2015;2015:535027.

[87] Minan A, Bosch A, Lasch P, Stammler M, Serra DO, Degrossi J, et al. Rapid identification of Burkholderia cepacia complex species including strains of the novel Taxon K, recovered from cystic fibrosis patients by intact cell MALDI-ToF mass spectrometry. Analyst 2009;134:1138−48.

[88] Niyompanich S, Jaresitthikunchai J, Srisanga K, Roytrakul S, Tungpradabkul S. Source-identifying biomarker ions between environmental and clinical Burkholderia pseudomallei using whole-cell matrix-assisted laser desorption/ionization time-of-flight mass spectrometry (MALDI-TOF MS). PLoS One 2014;9 e99160.

[89] Gomila M, Prince-Manzano C, Svensson-Stadler L, Busquets A, Erhard M, Martinez DL, et al. Genotypic and phenotypic applications for the differentiation and species-level identification of Achromobacter for clinical diagnoses. PLoS One 2014;9 e114356.

[90] Gherardi G, Creti R, Pompilio A, Di Bonaventura G. An overview of various typing methods for clinical epidemiology of the emerging pathogen Stenotrophomonas maltophilia. Diagn Microbiol Infect Dis 2015;81:219−26.

[91] Ostergaard C, Hansen SG, Moller JK. Rapid first-line discrimination of methicillin resistant Staphylococcus aureus strains using MALDI-TOF MS. Int J Med Microbiol 2015;305:838−47.

[92] Lasch P, Fleige C, Stammler M, Layer F, Nubel U, Witte W, et al. Insufficient discriminatory power of MALDI-TOF mass spectrometry for typing of Enterococcus faecium and Staphylococcus aureus isolates. J Microbiol Methods 2014;100:58−69.

[93] Zhang T, Ding J, Rao X, Yu J, Chu M, Ren W, et al. Analysis of methicillin-resistant Staphylococcus aureus major clonal lineages by matrix-assisted laser desorption ionization-time of flight mass spectrometry (MALDI-TOF MS). J Microbiol Methods 2015;117:122−7.

[94] Wolters M, Rohde H, Maier T, Belmar-Campos C, Franke G, Scherpe S, et al. MALDI-TOF MS fingerprinting allows for discrimination of major methicillin-resistant Staphylococcus aureus lineages. Int J Med Microbiol 2011;301:64−8.

[95] Josten M, Reif M, Szekat C, Al-Sabti N, Roemer T, Sparbier K, et al. Analysis of the matrix-assisted laser desorption ionization-time of flight mass spectrum of Staphylococcus aureus identifies mutations that allow differentiation of the main clonal lineages. J Clin Microbiol 2013;51:1809−17.

[96] Boggs SR, Cazares LH, Drake R. Characterization of a Staphylococcus aureus USA300 protein signature using matrix-assisted laser desorption/ionization time-of-flight mass spectrometry. J Med Microbiol 2012;61:640−4.

[97] Nakano S, Matsumura Y, Kato K, Yunoki T, Hotta G, Noguchi T, et al. Differentiation of vanA-positive Enterococcus faecium from vanA-negative E. faecium by matrix-assisted laser desorption/ionisation time-of-flight mass spectrometry. Int J Antimicrob Agents 2014;44:256−9.

[98] Griffin PM, Price GR, Schooneveldt JM, Schlebusch S, Tilse MH, Urbanski T, et al. Use of matrix-assisted laser desorption ionization-time of flight mass spectrometry to identify vancomycin-resistant enterococci and investigate the epidemiology of an outbreak. J Clin Microbiol 2012;50:2918−31.

[99] Freitas AR, Sousa C, Novais C, Silva L, Ramos H, Coque TM, et al. Rapid detection of high-risk Enterococcus faecium clones by matrix-assisted laser desorption ionization time-of-flight mass spectrometry. Diagn Microbiol Infect Dis 2017;87:299−307.

[100] Schlebusch S, Price GR, Gallagher RL, Horton-Szar V, Elbourne LD, Griffin P, et al. MALDI-TOF MS meets WGS in a VRE outbreak investigation. Eur J Clin Microbiol Infect Dis 2017;36:495−9.

[101] Nakano S, Matsumura Y, Ito Y, Fujisawa T, Chang B, Suga S, et al. Development and evaluation of MALDI-TOF MS-based serotyping for Streptococcus pneumoniae. Eur J Clin Microbiol Infect Dis 2015;34:2191−8.

[102] Pinto TC, Costa NS, Castro LF, Ribeiro RL, Botelho AC, Neves FP, et al. Potential of MALDI-TOF MS as an alternative approach for capsular typing Streptococcus pneumoniae isolates. Sci Rep 2017;7:45572.

[103] Lartigue MF, Kostrzewa M, Salloum M, Haguenoer E, Hery-Arnaud G, Domelier AS, et al. Rapid detection of "highly virulent" Group B Streptococcus ST-17 and emerging ST-1 clones by MALDI-TOF mass spectrometry. J Microbiol Methods 2011;86:262−5.

[104] Moura H, Woolfitt AR, Carvalho MG, Pavlopoulos A, Teixeira LM, Satten GA, et al. MALDI-TOF mass spectrometry as a tool for differentiation of invasive and noninvasive Streptococcus pyogenes isolates. FEMS Immunol Med Microbiol 2008;53:333−42.

[105] Mlaga KD, Dubourg G, Abat C, Chaudet H, Lotte L, Diene SM, et al. Using MALDI-TOF MS typing method to decipher outbreak: the case of Staphylococcus saprophyticus causing urinary tract infections (UTIs) in Marseille, France. Eur J Clin Microbiol Infect Dis 2017;36:2371−7.

[106] Sato J, Nakayama M, Tomita A, Sonoda T, Hasumi M, Miyamoto T. Evaluation of repetitive-PCR and matrix-assisted laser desorption ionization-time of flight mass spectrometry (MALDI-TOF MS) for rapid strain typing of Bacillus coagulans. PLoS One 2017;12 e0186327.

[107] Fernandez-No IC, Bohme K, Diaz-Bao M, Cepeda A, Barros-Velazquez J, Calo-Mata P. Characterisation and profiling of Bacillus subtilis, Bacillus cereus and Bacillus licheniformis by MALDI-TOF mass fingerprinting. Food Microbiol 2013;33:235−42.

[108] Ojima-Kato T, Yamamoto N, Takahashi H, Tamura H. Matrix-assisted laser desorption ionization-time of flight mass spectrometry (MALDI-TOF MS) can precisely discriminate the lineages of Listeria monocytogenes and species of Listeria. PLoS One 2016;11 e0159730.

[109] Kiyosuke M, Kibe Y, Oho M, Kusaba K, Shimono N, Hotta T, et al. Comparison of two types of matrix-assisted laser desorption/ionization time-of-flight mass spectrometer for the identification and typing of Clostridium difficile. J Med Microbiol 2015;64:1144−50.

[110] Nagy E, Urban E, Becker S, Kostrzewa M, Voros A, Hunyadkurti J, et al. MALDI-TOF MS fingerprinting facilitates rapid discrimination of phylotypes I, II and III of Propionibacterium acnes. Anaerobe 2013;20:20−6.

[111] Taniguchi T, Sekiya A, Higa M, Saeki Y, Umeki K, Okayama A, et al. Rapid identification and subtyping of Helicobacter cinaedi strains by intact-cell mass spectrometry profiling with the use of matrix-assisted laser desorption ionization-time of flight mass spectrometry. J Clin Microbiol 2014;52:95−102.

[112] Rizzardi K, Akerlund T. High molecular weight typing with MALDI-TOF MS—a novel method for rapid typing of Clostridium difficile. PLoS One 2015;10: e0122457.

[113] Kehrmann J, Wessel S, Murali R, Hampel A, Bange FC, Buer J, et al. Principal component analysis of MALDI TOF MS mass spectra separates M. abscessus (sensu stricto) from M. massiliense isolates. BMC Microbiol 2016;16:24.

[114] Suzuki H, Yoshida S, Yoshida A, Okuzumi K, Fukusima A, Hishinuma A. A novel cluster of Mycobacterium abscessus complex revealed by matrix-assisted laser desorption ionization-time-of-flight mass spectrometry (MALDI-TOF MS). Diagn Microbiol Infect Dis 2015;83:365−70.

[115] Verroken A, Bauraing C, Deplano A, Bogaerts P, Huang D, Wauters G, et al. Epidemiological investigation of a nosocomial outbreak of multidrug-resistant Corynebacterium striatum at one Belgian university hospital. Clin Microbiol Infect 2014;20:44−50.

[116] Trevino M, Areses P, Penalver MD, Cortizo S, Pardo F, del Molino ML, et al. Susceptibility trends of Bacteroides fragilis group and characterisation of carbapenemase-producing strains by automated REP-PCR and MALDI TOF. Anaerobe 2012;18:37−43.

[117] Rizzardi K, Wahab T, Jernberg C. Rapid subtyping of Yersinia enterocolitica by matrix-assisted laser desorption ionization-time of flight mass spectrometry (MALDI-TOF MS) for diagnostics and surveillance. J Clin Microbiol 2013;51:4200−3.

[118] Prakash A, Sharma C, Singh A, Kumar Singh P, Kumar A, Hagen F, et al. Evidence of genotypic diversity among Candida auris isolates by multilocus sequence typing, matrix-assisted laser desorption ionization time-of-flight mass spectrometry and amplified fragment length polymorphism. Clin Microbiol Infect 2016;22 277. e1−e9.

[119] Muller W, Hotzel H, Otto P, Karger A, Bettin B, Bocklisch H, et al. German Francisella tularensis isolates from European brown hares (Lepus europaeus) reveal genetic and phenotypic diversity. BMC Microbiol 2013;13:61.

[120] Quirino A, Pulcrano G, Rametti L, Puccio R, Marascio N, Catania MR, et al. Typing of Ochrobactrum anthropi clinical isolates using automated repetitive extragenic palindromic-polymerase chain reaction DNA fingerprinting and matrix-assisted laser desorption/ionization-time-of-flight mass spectrometry. BMC Microbiol 2014;14:74.

CHAPTER SIXTEEN

Application of MALDI-TOF in Parasitology

Juan de Dios Caballero and Oihane Martin

Department of Microbiology and Parasitology, Hospital Universitario Ramón y Cajal, Madrid, Spain

16.1 INTRODUCTION

16.1.1 Diagnostic tools in clinical parasitology

In recent years, matrix-assisted laser desorption/ionization−time-of-flight mass spectrometry (MALDI-TOF MS) technology has proved to be a useful tool in clinical microbiology, and it is successfully integrated in the diagnostic routine of many hospitals. Although this is certainly true for bacteriology and mycology, it is not applicable to parasitology yet [1−3].

Diagnosis of clinically important parasites is complex, requiring the detection and identification of a wide variety of organisms, from single-celled parasites to multicellular organisms, belonging to three main groups: protozoa, helminths, and ectoparasites [3]. Diagnostic tools commonly used in parasitology include microscopy, culture, serology, and molecular techniques. Resources in parasitology laboratories vary significantly among different hospitals; size, location, demographics, staff, and, particularly, the presence of units specialized in the diagnosis of tropical diseases, and determine the availability of diagnostic tools. In recent years, with the increase of international travel and migration, special attention has been drawn to the diagnosis of clinically important parasites and vector-borne diseases [4].

Despite the implementation of molecular techniques in many parasitology laboratories, diagnosis of parasitological diseases still relies heavily on microscopy. Microscopy is an affordable and highly specific technique that allows the multiple detection of parasites and remains the gold standard for many parasitological diseases [5−7]. It is routinely used in the diagnosis of intestinal protozoa and helminths (eggs and larvae), malaria

The Use of Mass Spectrometry Technology (MALDI-TOF) in Clinical Microbiology.
DOI: https://doi.org/10.1016/B978-0-12-814451-0.00016-2

(and other protozoa that can be found in blood), microfilariae, and for the identification of ectoparasites. In the latter, examination of morphological features is still the reference method used by entomologists and arachnologists [3,8], and the only technique available in most of the laboratories.

As a downside, this technique depends greatly on the microscopist's expertise, which is particularly difficult to acquire in the diagnosis of non-endemic parasitic diseases. Microscopy can be time-consuming, especially when complex staining techniques are required. This clashes with the current trend of centralization, personnel reduction, and work overload experienced in most of the laboratories. Microscopy sensitivity is also determined by the parasite load, requiring in some cases the collection and examination of multiple samples, which can be inconvenient for the patients. In a few cases, morphological identification alone might be insufficient for the differentiation between pathogenic and non-pathogenic species (e.g., *Entamoeba* spp.) [3,9,10]. In others, microscopy might not be useful at all, due to the parasite stage or the life cycle of the parasite itself. For that reason, culture, serology, and molecular techniques are widely used in the diagnosis of parasitological diseases along with microscopy [3,10]. Culture is often used to increase the parasite load (e.g., in leishmaniasis or trichomoniasis), but long incubation times are required, delaying the time of response. In case of intestinal parasites, cultures are technically demanding and are rarely used in clinical laboratories [3,10].

Rapid detection tests (RDTs), based on antigen or antibody detection by immunochromatography, are easy to perform and are available in most hospitals. Antigen detection tests are commonly used for the diagnosis of *Giardia lamblia* and *Cryptosporidium* spp. in fecal samples, with high sensitivity, although false positives and negatives have been reported [11]. In case of *Entamoeba histolytica* antigen detection, there are RDTs and ELISA plates commercially available, but they are not all equally useful for species-level identification, mainly due to cross-reactivity with non-pathogenic *Entamoeba dispar* [10]. RDTs for malaria are also widely used in the diagnosis of malaria, with acceptable sensitivity and specificity for falciparum malaria, although they are not useful for the identification of mixed infections or the determination of parasitemia [6]. Rapid immunochromatographic tests are also available for leishmaniasis, Chagas disease, and trichomoniasis diagnosis.

Serology is especially useful in the diagnosis of tissue-located protozoa and helminthic infections. It is a valuable tool in the diagnosis of amoebic abscesses, toxoplasmosis, Chagas disease (particularly for indeterminate and chronic stages), and leishmaniasis [10]. In many helminthic infections, microscopy has a limited value compared to serology: antibody detection often occurs prior to egg production, larvae or egg release can be intermittent (e.g., strongyloidiasis and schistosomiasis), and, in some cases, parasites are located in tissues, with no larvae or egg release (e.g., echinococcosis, cysticercosis, and toxocariasis). Nevertheless, positive predictive value of serology for disease diagnosis is limited due to the long-term presence of parasite-specific antibodies from past infections. In addition, specificity of serology might be low due to cross-reactivity between different parasite species and nonspecific results might be obtained in patients with some systemic diseases. False negatives, particularly in immunocompromised patients, may also occur [10].

In recent years, molecular methods such as PCR transformed the diagnostic routine of many parasitology laboratories, with many advantages over traditional microscopy. PCR is a highly sensitive and specific technique that allows multiple detection of parasites. In the case of intestinal parasites, there are currently commercial platforms available for the detection of intestinal protozoa in a single fecal sample. PCR is also useful for the detection of parasites that might be damaged by the lack of conservation, like *Dientamoeba fragilis*. In addition, molecular techniques allow the differentiation of pathogenic *E. histolytica* from nonpathogenic amoebae, the identification of *Blastocystis* spp. subtypes (described to be related to pathogenicity, transmission and response to treatment), and the detection and identification of *Cryptosporidium* spp. [3,9,10,12,13]. In the case of blood and tissue parasites, PCR is useful for the diagnosis of submicroscopic malaria (especially in semi-immune patients), the identification of mixed infections, and the detection of mutations that confer resistance to treatment [6,10]. It is also very helpful in leishmaniasis diagnosis and in the follow-up of Chagas disease [3,10,14]. It is routinely used in the diagnosis of keratitis by *Acanthamoeba* spp. along with culture and microscopy [15]. Finally, molecular detection of *Trichomonas vaginalis* is included in most of the molecular panels for the diagnosis of sexually transmitted infections.

Molecular tools for parasite detection, on the other hand, are not universally available, being restricted to big tertiary hospitals and reference laboratories. They are still expensive compared to microscopy, culture, and serology and require skilled personnel, equipment, and specific facilities. In addition, molecular methods usually applied to detect both alive and dead parasites, with the subsequent interpretation problems. Despite its high sensitivity, PCR can be inhibited by bile salts in feces, heme in blood, and by components commonly used in sample collection devices [16]. Finally, there is no consensus on the adequate methodology and/or housekeeping genes required for the identification of some parasites, such as *Blastocystis hominis* or some arthropod species [12,17].

16.1.2 The use of MALDI-TOF in parasitology: where are we now

MALDI-TOF MS technology consists in the co-crystallization of a given sample within an organic matrix solution, followed by its ionization by a laser beam. Ions are then desorbed from the matrix and accelerated at a fixed potential through a vacuum flight tube, in which a detection system located at its end measures ions' TOF. This information generates a characteristic spectrum for each analyzed sample that permits its identification by comparing it with a reference mass spectra database [18,19]. Although the development of MALDI-TOF MS in clinical parasitology has been much more limited compared to that observed in bacteriology or mycology, it is a promising tool for the identification of parasites and has gained popularity because it is a fast, easy-to-use, and cost effective approach that permits a high-throughput analysis by simultaneous processing of multiple samples [20,21].

16.2 IDENTIFICATION OF PARASITES BY MALDI-TOF MS

16.2.1 Protozoa

16.2.1.1 Intestinal protozoa

Proper differentiation of these parasites at the species level is becoming increasingly important, since there is growing evidence that these species could have differences in their clinical and epidemiological importance for humans [3]. There are few studies regarding the use of MALDI-TOF for species identification of intestinal protozoa. These studies, however, showed promising results demonstrating that MALDI-TOF MS technology could be used for protozoan species identification. This technique has also the ability of perform high-throughput analysis in a cost-effective manner, making it useful for epidemiological studies. As an example, correct identification of *Cryptosporidium* species is epidemiologically important, as it allows the identification of contamination sources, which is helpful for disease control in an outbreak setting [22−24].

Magnuson et al. [25] used MALDI-TOF MS technology to distinguish oocysts of *Cryptosporidium parvum* from *Cryptosporidium muris*, both obtained from immunocompromised infected mice. The purpose of this work was to contribute to the standardization of *C. parvum* oocysts production for research purposes. Oocysts from both species were washed with deionized water and subjected to a freeze-thawed process five times in order to break the cystic walls. The authors conclude that MALDI-TOF MS is able to discriminate oocysts from both species and that could be potentially used for quality control in oocysts production in a simple and rapid manner [25]. Later on, Glassmeyer et al. [26] developed an easier and improved methodology for the acquisition of *C. parvum* mass spectra, by washing oocysts three times with HPLC grade water and incubating them at least for 45 minutes with MALDI-TOF matrix (3,5-dimethoxy-4-hydroxy-cinnamic acid). They compared oocysts' mass spectra with those of intact sporozoites and found many common peaks between them. Therefore, they concluded that their sample preparation method was able to break oocysts' wall and that it allowed the obtention of quality mass spectra in a reproducible and easy manner [26]. Finally, Abal-Fabeiro et al. [22] identified *Cryptosporidium* species directly from 608 and 63 samples of human and bovine origin, respectively, using a method that combined genotyping and MALDI-TOF MS, and compared the results with the reference Sanger sequencing. This combined technique permitted a high-throughput genotyping of multiple samples simultaneously, and it correctly identified 90.9% of *Cryptosporidium* species with a sensitivity and specificity of 87.3% and 98%, respectively. The authors conclude that the combination between MALDI-TOF MS and genotyping is at least as effective as Sanger sequencing for the identification of *Cryptosporidium* species, being more cost-effective and less time-consuming when applied to a high number of samples [22].

MALDI-TOF MS has been also used for the subtyping of *Blastocystis* spp. Martiny et al. [12] constructed a reference database using six *Blastocystis* spp. isolates belonging to STs 1−4 and 8, and they used this library to analyze 19 *Blastocystis* spp. clinical isolates belonging to these five STs. The identification was correct in all cases when compared to the results of SSU-rDNA sequencing. Sample preparation for MALDI-TOF analysis required the obtention of solid axenic cultures, *Blastocystis* spp. colonies were either deposited directly on MALDI-TOF plate or were subjected to a prior ethanol—formic acid protein extraction procedure, being the results obtained by MALDI-TOF better with the latter. Additionally, MALDI-TOF MS was used to analyze protein extracts from nine liquid xenic cultures, but the identification obtained was only correct in two cultures in which *Blastocystis* spp. richness was higher. The authors conclude that MALDI-TOF MS is a promising and easy-to-use tool for *Blastocystis* spp. subtyping, but its use is limited in clinical laboratories at the present time due to the requirement of culturing axenically the parasite [12].

In addition, Calderaro et al. [9] reported the use of MALDI-TOF MS for the discrimination of pathogenic *E. histolytica* from commensal *E. dispar*. They obtained mass spectra of three *Entamoeba* reference species (*E. histolytica*, *E. dispar*, and *Entamoeba moshkovskii*), grown in axenic cultures, and compared them to 14 clinical strains (6 *E. histolytica* and 8 *E. dispar*) grown in monoxenic cultures with *Escherichia coli* ATCC 8739. A statistical analysis of mass to charge (m/z) peaks allowed the detection of two specific peaks for *E. histolytica* (8246 and 8303 Da) and three for *E. dispar* (4714, 5541, and 8207 Da). Moreover, MALDI-TOF MS analysis was able to detect those peaks directly in two fecal samples after 12—24 hours of incubation in Robinson's liquid medium without sera. The authors concluded that MALDI-TOF MS is able to discriminate between *E. histolytica* and *E. dispar* grown in monoxenic cultures and that it could potentially differentiate parasite species directly on fecal samples [9].

Finally, Villegas et al. [27] applied MALDI-TOF MS technology for the differentiation of *G. lamblia* and *Giardia muris*. The authors obtained intact cysts from experimentally infected rodents and, after a washing procedure, incubated them at least for 60 minutes at room temperature with matrix (3,5-dimethoxy-4-hydroxy-cinnamic acid) before spotting the samples on MALDI plates. Analysis of the obtained spectra showed differential peaks for both species, which demonstrated the ability of MALDI-TOF to distinguish between them. The authors concluded that MALDI-TOF has the potential to become a high-throughput technique to be used in epidemiological studies [27].

16.2.1.2 Blood and tissue protozoa
16.2.1.2.1 Leishmania:
A proper identification of *Leishmania* species involved in human infections is necessary for the study of disease epidemiology and for the

implementation of control measures in the absence of an effective vaccine [3]. In addition, clinical presentation (cutaneous, mucocutaneous, and visceral leishmaniasis) and disease prognosis depend on the species involved in the infection, as well as the selection of the most appropriate therapeutic regimen [28—31]. Gold standard identification procedures (isoenzyme analysis) are costly, difficult to perform, and labor-intensive being its use limited to reference laboratories [32].

There are increasing reports on the use of MALDI-TOF for *Leishmania* species identification from specimens obtained from clinical samples of infected patients [33—35]. In all reports, having a positive in vitro culture from clinical samples was a necessary previous step for the identification of *Leishmania* spp. Mouri et al. [35] were the first to describe the use of MALDI-TOF for the identification of protozoa. Their analysis of *Leishmania* species mass spectra obtained from clinical samples and reference strains showed that MALDI-TOF had at least the same sensitivity and specificity as the reference method for the identification of *Leishmania* spp., being performed in a few hours after culture positivity (80% were positive after 10 days). In addition, they developed a database-independent algorithm able to differentiate the main *Leishmania* species involved in human infection, including those belonging to the *Viannia* subgenus, which are associated with metastatic mucocutaneous leishmaniasis [35]. Cassagne et al. [33] and Culha et al. [34] compared the mass spectra obtained from *Leishmania* spp. clinical isolates with an in-house database built with spectra from well-characterized *Leishmania* species. Although this prior step is time-consuming and requires a representative and well-characterized *Leishmania* species collection, they were also able to correctly identify at the species level by MALDI-TOF the majority of the *Leishmania* clinical isolates [33,34]. Cassagne et al. [33] described a simplified method of samples' preparation, in which promastigotes were washed and concentrated in saline solution and deposited directly onto the plate without a protein extraction step [33].

16.2.1.2.2 *Trypanosoma brucei* and *Trypanosoma cruzi:*
There is few information about the use of MALDI-TOF MS for the identification of other blood and tissue trypanosomatids other than *Leishmania* spp. Avila et al. [36] described a method for MALDI-TOF MS identification of isolated trypanosomatids belonging to the species *T. cruzi* and *T. brucei* [36]. Parasites were grown in vitro and directly deposited onto a MALDI plate before mass spectra acquisition and

analysis. Their method was able to discriminate both species, and it could also differentiate *T. cruzi* life stages. It was not sensitive enough, however, to identify the different *T. cruzi* discrete typing units (DTUs). The authors conclude that these promising results could be a prelude to the development of a MS method for direct parasite detection and identification in biological fluids [36].

16.2.1.2.3 *Plasmodium* spp. and *Babesia* spp.

There are also initial but promising reports about the use of MALDI-TOF MS for the diagnosis of the *Apicomplexa* blood parasites *Plasmodium* spp. and *Babesia* spp. MALDI-TOF MS has been used by Demirev et al. [37] and Scholl et al. [38] to detect *Plasmodium* spp. in infected mice using an ultraviolet laser beam. After a simple and rapid sample preparation method, they were able to detect parasites inside the erythrocytes by the detection of individual heme molecules resulted from the ionization of hemozoin. This method showed a sensitivity of less than 10 parasites per microliter of blood, that is, higher than that of conventional microscopy. Malaria speciation was not, however, achievable by this method, and specificity against confounding factors as other infectious diseases or blood genetic disorders was not assessed. The authors conclude that this method could be an easy and rapid screening assay to select positive malaria blood samples for microscopic examination [37,38]. MALDI-TOF MS has demonstrated also its usefulness for the rapid detection of malaria-infected *Anopheles* mosquitoes in epidemiological surveys [39].

The detection of *Babesia* spp. in blood samples by microscopy is difficult due to the low levels of parasitemia produced by this microorganism. Adaszek et al. [40] showed that MALDI-TOF MS was able to discriminate between *Babesia canis* infected and noninfected dogs by the specific detection of a 51−52 kDa protein fraction in the infected group's sera. Results using this approach were concordant with those of PCR, being MALDI-TOF MS a more rapid and cost-effective technique [40]. In a later study, Dzięgiel et al. [41] demonstrated the specificity of MALDI-TOF MS for *B. canis* diagnosis in infected dogs against infections caused by *Anaplasma phagocytophilum* and *Borrelia burgdorferi* [41].

16.2.1.3 Other protozoa

MALDI-TOF MS has been also evaluated for the identification of *Acanthamoeba* spp. genotypes, as only some of the 20 described genotypes can cause keratitis and encephalitis in humans [42,43]. Other free-living amoeba that cause human disease is *Naegleria fowleri*, which is the only one of at least 30 species described for its genus that causes primary human meningoencephalitis [42,43]. Del Chierico et al. [15] developed a MALDI-TOF MS—based method for the identification and genotyping of *Acanthamoeba* spp. clinical strains. In the study, the mass spectra of 15 *Acanthamoeba* clinical strains at different incubation times from agar plates with heat-inactivated *E. coli* extract were analyzed after a simple protein extraction step with ethanol and 70% formic acid was performed. Genotyping of the strains was carried out by 18S rDNA sequencing, which served also for the construction of a neighbor-joining phylogenetic tree. Spectra obtained by MALDI-TOF MS allowed to differentiate all these genotypes, and to group them in the same clusters as those obtained by phylogenetic analysis. The authors conclude that MALDI-TOF MS is a powerful, easy-to-use, and rapid platform for the identification of *Acanthamoeba* spp. clinical isolates, and has also the potential to be used for direct diagnosis of acanthamoebiasis from clinical samples [15]. Same conclusions were obtained by Moura et al. [44] after the MALDI-TOF MS analysis of 18 *Naegleria fowleri* strains belonging to three genotypes obtained from both clinical and environmental samples [44].

16.2.2 Helminths

The non-taxonomic term "helminths" comprises flat and round worms belonging to the *phyla Platyhelminthes* and *Nematoda*. A few species of a third *phylum*, *Acanthocephala*, can also cause disease in humans (https://www.cdc.gov/dpdx/acanthocephaliasis/index.html). The identification of the members of these *phyla* requires a high-degree of expertise, is time consuming, and usually limited to adult specimens and not to eggs or developmental stages [45]. There are few reports on the use of MALDI-TOF for nematode species identification and most of them refer to plant pathogens, for which a rapid and reliable identification is crucial for the implementation of disease control measures [46]. However, the successful identification of *Trichinella* species by MALDI-TOF MS has been reported by Mayer-Scholl et al. [47], after the construction of a reference database with mass spectra from the nine described species of the genus. A simple extraction method with formic acid/acetonitrile plus sonication with silica beads was required prior to MALDI-TOF identification, as well as an amount of 3—5 larvae in order to obtain high-quality protein spectra. Identification results were comparable to those obtained by reference molecular methods, but the process lasted less than 1 hour. In addition, MALDI-TOF identification was not affected by specimens' storage procedures, like freezing or alcohol conservation [47].

MALDI-TOF MS technology has also been applied in helminthic diseases for research purposes. MS, in particular, has shown to be a useful tool for the study of the proteome of parasites, that could be used for the search of serum biomarkers of infection for a prompt disease diagnosis or for the identification of potential antigens for vaccine or RDT development [48–60]. However, MALDI-TOF MS devices used in these studies are not usually available for conventional clinical laboratories. In addition, these studies often require a complex pretreatment of the samples, consisting generally in the separation of sample proteins, usually by chromatography or electrophoresis, but also by magnetic beads and monoclonal antibodies [61]. Nevertheless, these pioneering studies could pave the way for a future implementation of MALDI-TOF MS technology for clinical diagnosis of helminthic diseases.

16.2.3 Arthropods

Phylum Arthropoda comprises more than 1 million species, including human ectoparasites and/or vectors of important viral, bacterial, or protozoal diseases such as dengue, Lyme's disease, or malaria, among others [17]. The incidence of many arthropod-borne diseases has been on the rise in recent years, as well as vectors' distribution areas, due to climatic changes, alteration of the environment, and human or animal migrations [62]. Examples of these phenomena are the rapid expansion of the West Nile Virus in the United States or the emergence of Crimean–Congo fever in Europe [63,64]. Rapid and accurate identification of arthropod vectors is crucial for the control of these diseases [17]. In addition, proper identification of ectoparasites in clinical laboratories could speed up the administration of a correct treatment, due to the specificity of some vector species for particular pathogens [17].

An expert entomologist, however, is required for morphological identification of arthropod species using identification keys. However, the number of experienced entomologists is declining in recent years, and their work is labor-intensive and time-consuming, especially when the number of specimens to analyze is high [17]. Proper identification of incomplete, immature, or engorged specimens is often difficult, as well as the differentiation of nearly identical sibling or cryptic species with different epidemiological impact [17]. MALDI-TOF MS is, on the other hand, a rapid, easy, and reliably high-throughput technique that could be used for arthropod identification in epidemiological studies or in the clinical setting.

There are several studies regarding the use of MALDI-TOF MS for the identification of mosquitoes [65,66], midges [67], tsetse flies [68], ticks [8,69−73], sand flies [74,75], and fleas [76]. All of them reported a successful identification of arthropod species by MALDI-TOF when compared to conventional and/or molecular identification, being the construction of a reference mass spectra database a prior step for the process in almost all studies. With the help of a complete library, MALDI-TOF MS was also able in some studies to differentiate between sister or cryptic arthropod species [70,77,78], to distinguish different life stages within species [69] and even to speciate the genus *Aedes* by the analysis of mass spectra obtained from egg samples [65]. Interestingly, Yssouf et al. [71,72] demonstrated the ability of MALDI-TOF analysis to discriminate between *Rickettsia* spp. infected and noninfected ticks, which was also capable to correctly identify this pathogen to the species level.

The construction of a library is a simple process, requiring the homogenization of specimens and a protein extraction step with 70% formic acid and acetonitrile [17]. There are, however, two crucial issues to consider before constructing a database to obtain quality and reproducible mass spectra: the conservation methods employed to preserve the specimens (especially in field studies) and the adequate choice of the body parts to analyze [17]. Fresh and frozen specimens are the best choice for mass spectra acquisition, whereas preservation with 70% ethanol has shown to reduce its quality and reproducibility in some studies [17,67,76]. Dry preservation in silica gel, however, does not seem to affect the acquisition of mass spectra [78]. The selection of body parts is also crucial because protein profiling varies according to the different body parts [17]. The inclusion of the abdomen is generally not recommended, especially in haematophagous species, because it decreases the reproducibility and quality of MALDI-TOF analysis [17]. Thus, it is important to evaluate the effect of preservation methods and chosen body parts for analysis on mass spectra acquisition, before the completion of the database [17]. Finally, a limiting factor on the extensive use of MALDI-TOF MS for arthropod identification is the inability to use this technology directly on field studies, although advances in that way are under development [3].

16.3 IMPLEMENTATION OF MALDI-TOF IN THE PARASITOLOGY LABORATORY: PROSPECTS AND LIMITATIONS

The development of MALDI-TOF MS in clinical parasitology has been much more limited compared to that observed in bacteriology or mycology, and it is not currently used for the diagnosis of parasitic diseases and/or parasite species identification in clinical laboratories. There are multiple reasons for this lack of MS application in clinical parasitology: first, the variety of life-forms present in many parasites (cysts, trophozoites, eggs, etc.) from which different protein profiles can be obtained by MS; second, the complexity of parasite life cycles that sometimes make impossible to obtain a parasite specimen from infected humans; and finally, the requirement for an in vitro culture of parasites to obtain a mass spectra for its identification, which is difficult, time-consuming, and is not performed in the majority of clinical laboratories [3]. High concentration of parasites is required for protein extraction and species identification [9,33]. This is strongly dependent on the initial parasite load of the clinical sample and the incubation time, that may range from a few days up to weeks.

Successful cultivation of parasites often requires co-cultivation with other organisms such as *E. coli* [12]. Spectra from *E. coli* or other accompanying microorganisms are also obtained in the analysis and need to be ruled out [9]. MS identification of parasites obtained by xenic cultures is less accurate than the one obtained from axenic cultures, and it depends on the ratio parasites/bacteria or fungus used for cultivation. This limitation is also true for the analysis of clinical samples, in which other microorganism are present as part of the microbiota, such as in fecal samples [12].

It is probably because of the above-mentioned reasons that the different commercial software employed for MALDI-TOF MS in clinical laboratories, such as Bruker Biotyper (Bruker Daltonics, Bremen, Germany) or VITEK MS (BioMérieux, Nürtingen, Germany) among others, have not included any parasite reference spectra in their database [79]. Fortunately, most of the commercial softwares include the possibility to create additional reference spectra to include those of the missing species [79].

Despite these limitations, there are increasing reports of the use of MALDI-TOF for the identification of parasite clinical and field-collected isolates, in which this technology has shown to be equal to molecular tools in terms of accuracy and reliability, being in addition cost-effective despite the initial cost of the MS device [79]. In addition, there are other reports showing the potential of MALDI-TOF MS analysis for diagnosing parasitic infections directly from clinical samples, identifying parasitic biomarkers in infected patients, and searching for parasite antigens for vaccine or RDTs development [3,79]. The key would be to identify those circumstances in which the use of MALDI-TOF could be cost-effective, which would be strongly determined by the resources available in each laboratory. The use of MALDI-TOF for the diagnosis of parasitic diseases is still restricted to research laboratories due to technical requirements, although advances in MS technology could solve this limitation in the future [80]. Lack of technical resources is also true for MALDI-TOF identification of clinically relevant protozoa, as there is still a need of culturing parasites in vitro [3]. Identification of arthropod ectoparasites for epidemiological studies is, at the moment, the most promising application of MALDI-TOF MS technology when trained staff is not available and there are high amounts of specimens to analyze, solving additional problems as the identification of immature or incomplete specimens or cryptic species [17]. This application, however, seems to be useful only in centers performing epidemiological studies, as ectoparasites are not frequently collected in clinical laboratories, in which spending efforts in constructing reference databases could not be worthwhile.

To conclude, MALDI-TOF MS is a promising tool for parasite identification, currently applicable in reference laboratories. However, there are some issues to be resolved before its implementation as a routine diagnostic tool in clinical laboratories.

REFERENCES

[1] Cassagne C, Normand AC, L'Ollivier C, Ranque S, Piarroux R. Performance of MALDI-TOF MS platforms for fungal identification. Mycoses 2016;59:678—90. Available from: https://doi.org/10.1111/myc.12506.

[2] Sandalakis V, Goniotakis I, Vranakis I, Chochlakis D, Psaroulaki A. Use of MALDI-TOF mass spectrometry in the battle against bacterial infectious diseases: recent achievements and future perspectives. Expert Rev Proteomics 2017;14:253—67. Available from: https://doi.org/10.1080/14789450.2017.1282825.

[3] Singhal N, Kumar M, Virdi JS. MALDI-TOF MS in clinical parasitology: applications, constraints and prospects. Parasitology 2016;143:1491—500. Available from: https://doi.org/10.1017/S0031182016001189.

[4] Field V, Gautret P, Schlagenhauf P, Burchard GD, Caumes E, Jensenius M, et al. Travel and migration associated infectious diseases morbidity in Europe, 2008. BMC Infect Dis 2010;10:330.

[5] McHardy IH, Wu M, Shimizu-Cohen R, Couturier MR, Humphries RM. Detection of intestinal protozoa in the clinical laboratory. J Clin Microbiol 2014;52:712−20. Available from: https://doi.org/10.1128/JCM.02877-13.

[6] Moody A. Rapid diagnostic tests for malaria parasites. Clin Microbiol Rev 2002;15:66−78. Available from: https://doi.org/10.1128/CMR.15.1.66-78.2002.

[7] van Lieshout L, Roestenberg M. Clinical consequences of new diagnostic tools for intestinal parasites. Clin Microbiol Infect 2015;21:520−8. Available from: https://doi.org/10.1016/j.cmi.2015.03.015.

[8] Kumsa B, Laroche M, Almeras L, Mediannikov O, Raoult D, Parola P. Morphological, molecular and MALDI-TOF mass spectrometry identification of ixodid tick species collected in Oromia, Ethiopia. Parasitol Res 2016;115:4199−210. Available from: https://doi.org/10.1007/s00436-016-5197-9.

[9] Calderaro A, Piergianni M, Buttrini M, Montecchini S, Piccolo G, Gorrini C, et al. MALDI-TOF mass spectrometry for the detection and differentiation of *Entamoeba histolytica* and *Entamoeba dispar*. PLoS One 2015;10:e0122448. Available from: https://doi.org/10.1371/journal.pone.0122448.

[10] Theel ES, Pritt BS. Parasites. Microbiol Spectr 2016;4. Available from: https://doi.org/10.1128/microbiolspec.DMIH2-0013-2015.

[11] Weitzel T, Dittrich S, Möhl I, Adusu E, Jelinek T. Evaluation of seven commercial antigen detection tests for *Giardia* and *Cryptosporidium* in stool samples. Clin Microbiol Infect 2006;12:656−9. Available from: https://doi.org/10.1111/j.1469-0691.2006.01457.x.

[12] Martiny D, Bart A, Vandenberg O, Verhaar N, Wentink-Bonnema E, Moens C, et al. Subtype determination of *Blastocystis* isolates by matrix-assisted laser desorption/ionisation time-of-flight mass spectrometry (MALDI-TOF MS). Eur J Clin Microbiol Infect Dis 2014;33:529−36. Available from: https://doi.org/10.1007/s10096-013-1980-z.

[13] Rolando RF, Silva Sd, Peralta RH, Silva AJ, Cunha Fde S, Bello AR, et al. Detection and differentiation of *Cryptosporidium* by real-time polymerase chain reaction in stool samples from patients in Rio de Janeiro, Brazil. Mem Inst Oswaldo Cruz 2012;107:476−9.

[14] Britto CC. Usefulness of PCR-based assays to assess drug efficacy in Chagas disease chemotherapy: value and limitations. Mem Inst Oswaldo Cruz 2009;104:122−35.

[15] Del Chierico F, Di Cave D, Accardi C, Santoro M, Masotti A, D'Alfonso R, et al. Identification and typing of free-living *Acanthamoeba* spp. by MALDI-TOF MS Biotyper. Exp Parasitol 2016;170:82−9. Available from: https://doi.org/10.1016/j.exppara.2016.09.007.

[16] Buckwalter SP, Sloan LM, Cunningham SA, Espy MJ, Uhl JR, Jones MF, et al. Inhibition controls for qualitative real-time PCR assays: are they necessary for all specimen matrices? J Clin Microbiol 2014;52:2139−43. Available from: https://doi.org/10.1128/JCM.03389-13.

[17] Yssouf A, Almeras L, Raoult D, Parola P. Emerging tools for identification of arthropod vectors. Future Microbiol 2016;11:549−66. Available from: https://doi.org/10.2217/fmb.16.5.

[18] Karlsson R, Gonzales-Siles L, Boulund F, Svensson-Stadler L, Skovbjerg S, Karlsson A, et al. Proteotyping: proteomic characterization, classification and identification of microorganisms—a prospectus. Syst Appl Microbiol 2015;38:246−57. Available from: https://doi.org/10.1016/j.syapm.2015.03.006.

[19] Welker M, Moore ER. Applications of whole-cell matrix-assisted laser-desorption/ionization time-of-flight mass spectrometry in systematic microbiology. Syst Appl Microbiol 2011;34:2−11. Available from: https://doi.org/10.1016/j.syapm.2010.11.013.

[20] Ge MC, Kuo AJ, Liu KL, Wen YH, Chia JH, Chang PY, et al. Routine identification of microorganisms by matrix-assisted laser desorption ionization time-of-flight mass spectrometry: success rate, economic analysis, and clinical outcome. J Microbiol Immunol Infect 2017;50:662−8. Available from: https://doi.org/10.1016/j.jmii.2016.06.002.

[21] Tran A, Alby K, Kerr A, Jones M, Gilligan PH. Cost savings realized by implementation of routine microbiological identification by matrix-assisted laser desorption ionization-time of flight mass spectrometry. J Clin Microbiol 2015;53:2473−9. Available from: https://doi.org/10.1128/JCM.00833-15.

[22] Abal-Fabeiro JL, Maside X, Llovo J, Bello X, Torres M, Treviño M, et al. High-throughput genotyping assay for the large-scale genetic characterization of Cryptosporidium parasites from human and bovine samples. Parasitology 2014;141:491−500. Available from: https://doi.org/10.1017/S0031182013001807.

[23] Chalmers RM. Cryptosporidium: from laboratory diagnosis to surveillance and outbreaks. Parasite 2008;15:372−8. Available from: https://doi.org/10.1051/parasite/2008153372.

[24] Jex AR, Gasser RB. Genetic richness and diversity in Cryptosporidium hominis and C. parvum reveals major knowledge gaps and a need for the application of "next generation" technologies—research review. Biotechnol Adv 2010;28:17−26. Available from: https://doi.org/10.1016/j.biotechadv.2009.08.003.

[25] Magnuson ML, Owens JH, Kelty CA. Characterization of Cryptosporidium parvum by matrix-assisted laser desorption ionization-time of flight mass spectrometry. Appl Environ Microbiol 2000;66:4720−4. Available from: https://doi.org/10.1128/AEM.66.11.4720-4724.2000.

[26] Glassmeyer ST, Ware MW, Schaefer FW, Shoemaker JA, Kryak DD. An improved method for the analysis of Cryptosporidium parvum oocysts by matrix-assisted laser desorption/ionization time of flight mass spectrometry. J Eukaryot Microbiol 2007;54:479−81. Available from: https://doi.org/10.1111/j.1550-7408.2007.00287.x.

[27] Villegas EN, Glassmeyer ST, Ware MW, Hayes SL, Schaefer FW. Matrix-assisted laser desorption/ionization time-of-flight mass spectrometry-based analysis of Giardia lamblia and Giardia muris. J Eukaryot Microbiol 2006;53:S179−81. Available from: https://doi.org/10.1111/j.1550-7408.2006.00223.x.

[28] Arevalo J, Ramirez L, Adaui V, Zimic M, Tulliano G, Miranda-Verastegui C, et al. Influence of Leishmania (Viannia) species on the response to antimonial treatment in patients with American tegumentary leishmaniasis. J Infect Dis 2007;195:1846−51. Available from: https://doi.org/10.1086/518041.

[29] Modabber F, Buffet PA, Torreele E, Milon G, Croft SL. Consultative meeting to develop a strategy for treatment of cutaneous leishmaniasis. Institute Pasteur, Paris. 13−15 June, 2006. Kinetoplastid Biol Dis 2007;6:3. Available from: https://doi.org/10.1186/1475-9292-6-3.

[30] Navin TR, Arana BA, Arana FE, Berman JD, Chajon JF. Placebo-controlled clinical trial of sodium stibogluconate (Pentostam) versus ketoconazole for treating cutaneous leishmaniasis in Guatemala. J Infect Dis 1992;165:528−34.

[31] Neal RA, Allen S, McCoy N, Olliaro P, Croft SL. The sensitivity of Leishmania species to aminosidine. J Antimicrob Chemother 1995;35:577−84.

[32] Pratlong F, Dereure J, Ravel C, Lami P, Balard Y, Serres G, et al. Geographical distribution and epidemiological features of Old World cutaneous leishmaniasis foci, based on the isoenzyme analysis of 1048 strains. Trop Med Int Health 2009;14:1071−85. Available from: https://doi.org/10.1111/j.1365-3156.2009.02336.x.

[33] Cassagne C, Pratlong F, Jeddi F, Benikhlef R, Aoun K, Normand AC, et al. Identification of Leishmania at the species level with matrix-assisted laser desorption

ionization time-of-flight mass spectrometry. Clin Microbiol Infect 2014;20:551−7. Available from: https://doi.org/10.1111/1469-0691.12387.

[34] Culha G, Akyar I, Zeyrek FY, Kurt Ö, Gündüz C, Töz SÖ, et al. Leishmaniasis in Turkey: determination of *Leishmania* species by matrix-assisted laser desorption ionization time-of-flight mass spectrometry (MALDI-TOF MS). Iran J Parasitol 2014;9:239−48.

[35] Mouri O, Morizot G, Van der Auwera G, Ravel C, Passet M, Chartrel N, et al. Easy identification of *Leishmania* species by mass spectrometry. PLoS Negl Trop Dis 2014;8:e2841. Available from: https://doi.org/10.1371/journal.pntd.0002841.

[36] Avila CC, Almeida FG, Palmisano G. Direct identification of trypanosomatids by matrix-assisted laser desorption ionization−time of flight mass spectrometry (DIT MALDI-TOF MS). J Mass Spectrom 2016;549−57. Available from: https://doi.org/10.1002/jms.3763.

[37] Demirev PA, Feldman AB, Kongkasuriyachai D, Scholl P, Sullivan DJ, Kumar N. Detection of malaria parasites in blood by laser desorption mass spectrometry. Anal Chem 2002;74:3262−6.

[38] Scholl PF, Kongkasuriyachai D, Demirev PA, Feldman AB, Lin JS, Sullivan DJ, et al. Rapid detection of malaria infection in vivo by laser desorption mass spectrometry. Am J Trop Med Hyg 2004;71:546−51.

[39] Laroche M, Almeras L, Pecchi E, Bechah Y, Raoult D, Viola A, et al. MALDI-TOF MS as an innovative tool for detection of *Plasmodium* parasites in *Anopheles* mosquitoes. Malar J 2017;16:5. Available from: https://doi.org/10.1186/s12936-016-1657-z.

[40] Adaszek Ł, Banach T, Bartnicki M, Winiarczyk D, Łyp P, Winiarczyk S. Application the mass spectrometry MALDI-TOF technique for detection of *Babesia canis canis* infection in dogs. Parasitol Res 2014;113:4293−5. Available from: https://doi.org/10.1007/s00436-014-4124-1.

[41] Dzięgiel B, Adaszek Ł, Banach T, Winiarczyk S. Specificity of mass spectrometry (MALDI-TOF) in the diagnosis of *Babesia canis* regarding to other canine vector-borne diseases. Ann Parasitol 2016;62:101−5. Available from: https://doi.org/10.17420/ap6202.39.

[42] Visvesvara GS. Infections with free-living amebae. Handb Clin Neurol 2013;114:153−68. Available from: https://doi.org/10.1016/B978-0-444-53490-3.00010-8.

[43] Visvesvara GS, Moura H, Schuster FL. Pathogenic and opportunistic free-living amoebae: *Acanthamoeba* spp., *Balamuthia mandrillaris*, *Naegleria fowleri*, and *Sappinia diploidea*. FEMS Immunol Med Microbiol 2007;50:1−26. Available from: https://doi.org/10.1111/j.1574-695X.2007.00232.x.

[44] Moura H, Izquierdo F, Woolfitt AR, Wagner G, Pinto T, Del Aguila C, et al. Detection of biomarkers of pathogenic *Naegleria fowleri* through mass spectrometry and proteomics. J Eukaryot Microbiol 2015;62:12−20. Available from: https://doi.org/10.1111/jeu.12178.

[45] Bredtmann CM, Krücken J, Murugaiyan J, Kuzmina T, von Samson-Himmelstjerna G. Nematode species identification—current status, challenges and future perspectives for cyathostomins. Front Cell Infect Microbiol 2017;7:1−8. Available from: https://doi.org/10.3389/fcimb.2017.00283.

[46] Ahmad F, Babalola OO, Tak HI. Potential of MALDI-TOF mass spectrometry as a rapid detection technique in plant pathology: identification of plant-associated microorganisms. Anal Bioanal Chem 2012;404:1247−55. Available from: https://doi.org/10.1007/s00216-012-6091-7.

[47] Mayer-Scholl A, Murugaiyan J, Neumann J, Bahn P, Reckinger S, Nöckler K. Rapid identification of the foodborne pathogen *Trichinella* spp. by matrix-assisted

laser desorption/ionization mass spectrometry. PLoS One 2016;11:e0152062. Available from: https://doi.org/10.1371/journal.pone.0152062.

[48] Garcia-Campos A, Ravidà A, Nguyen DL, Cwiklinski K, Dalton JP, Hokke CH, et al. Tegument glycoproteins and cathepsins of newly excysted juvenile *Fasciola hepatica* carry mannosidic and paucimannosidic N-glycans. PLoS Negl Trop Dis 2016;10:e0004688. Available from: https://doi.org/10.1371/journal.pntd.0004688.

[49] Haçariz O, Baykal AT, Akgün M, Kavak P, Sağiroğlu MŞ, Sayers GP. Generating a detailed protein profile of *Fasciola hepatica* during the chronic stage of infection in cattle. Proteomics 2014;14:1519—30. Available from: https://doi.org/10.1002/pmic.201400012.

[50] Huang Y, Li W, Liu K, Xiong C, Cao P, Tao J. New detection method in experimental mice for schistosomiasis: ClinProTool and matrix-assisted laser desorption/ionization time-of-flight mass spectrometry. Parasitol Res 2016;115:4173—81. Available from: https://doi.org/10.1007/s00436-016-5193-0.

[51] Kardoush MI, Ward BJ, Ndao M. Identification of candidate serum biomarkers for *Schistosoma mansoni* infected mice using multiple proteomic platforms. PLoS One 2016;11:e0154465. Available from: https://doi.org/10.1371/journal.pone.0154465.

[52] Liu RD, Cui J, Wang L, Long SR, Zhang X, Liu MY, et al. Identification of surface proteins of *Trichinella spiralis* muscle larvae using immunoproteomics. Trop Biomed 2014;31:579—91. Available from: https://doi.org/10.1186/1756-3305-7-40.

[53] Liu RD, Qi X, Sun GG, Jiang P, Zhang X, Wang LA, et al. Proteomic analysis of *Trichinella spiralis* adult worm excretory-secretory proteins recognized by early infection sera. Vet Parasitol 2016;231:43—6. Available from: https://doi.org/10.1016/j.vetpar.2016.10.008.

[54] Rioux MC, Carmona C, Acosta D, Ward B, Ndao M, Gibbs BF, et al. Discovery and validation of serum biomarkers expressed over the first twelve weeks of *Fasciola hepatica* infection in sheep. Int J Parasitol 2008;38:123—36. Available from: https://doi.org/10.1016/j.ijpara.2007.07.017.

[55] Robijn ML, Planken J, Kornelis D, Hokke CH, Deelder AM. Mass spectrometric detection of urinary oligosaccharides as markers of *Schistosoma mansoni* infection. Trans R Soc Trop Med Hyg 2008;102:79—83. Available from: https://doi.org/10.1016/j.trstmh.2007.09.017.

[56] Balog CI, Alexandrov T, Derks RJ, Hensbergen PJ, van Dam GJ, Tukahebwa EM, et al. The feasibility of MS and advanced data processing for monitoring *Schistosoma mansoni* infection. Proteomics Clin Appl 2010;4:499—510. Available from: https://doi.org/10.1002/prca.200900158.

[57] Wang X, Chen W, Li X, Zhou C, Deng C, Lv X, et al. Identification and molecular characterization of a novel signaling molecule 14-3-3 epsilon in *Clonorchis sinensis* excretory/secretory products. Parasitol Res 2012;110:1411—20. Available from: https://doi.org/10.1007/s00436-011-2642-7.

[58] Boamah D, Kikuchi M, Huy NT, Okamoto K, Chen H, Ayi I, et al. Immunoproteomics identification of major IgE and IgG4 reactive *Schistosoma japonicum* adult worm antigens using chronically infected human plasma. Trop Med Health 2012;40:89—102. Available from: https://doi.org/10.2149/tmh.2012-16.

[59] Wang Y, Xiao D, Shen Y, Han X, Zhao F, Li X, et al. Proteomic analysis of the excretory/secretory products and antigenic proteins of *Echinococcus granulosus* adult worms from infected dogs. BMC Vet Res 2015;11:119. Available from: https://doi.org/10.1186/s12917-015-0423-8.

[60] Weinkopff T, Atwood JA, Punkosdy GA, Moss D, Weatherly DB, Orlando R, et al. Identification of antigenic *Brugia* adult worm proteins by peptide mass fingerprinting. J Parasitol 2009;95:1429—35. Available from: https://doi.org/10.1645/GE-2083.1.

[61] Cheng K, Chui H, Domish L, Hernandez D, Wang G. Recent development of mass spectrometry and proteomics applications in identification and typing of bacteria. Proteomics Clin Appl 2016;10:346−57. Available from: https://doi.org/10.1002/prca.201500086.

[62] Smolinski M, Hamburg M, Lederberg J, editors. Microbial threats to health emergence, detection, and response. Washington (DC): Institute of Medicine (Us) Committee on Emerging Microbial Threats to Health in the 21st Century, National Academies Press (US); 2003.

[63] Maltezou HC, Andonova L, Andraghetti R, Bouloy M, Ergonul O, Jongejan F, et al. Crimean-Congo hemorrhagic fever in Europe: current situation calls for preparedness. Euro Surveill 2010;15:19504.

[64] Wimberly MC, Lamsal A, Giacomo P, Chuang TW. Regional variation of climatic influences on West Nile virus outbreaks in the United States. Am J Trop Med Hyg 2014;91:677−84. Available from: https://doi.org/10.4269/ajtmh.14-0239.

[65] Schaffner F, Kaufmann C, Pflüger V, Mathis A. Rapid protein profiling facilitates surveillance of invasive mosquito species. Parasit Vectors 2014;7:142. Available from: https://doi.org/10.1186/1756-3305-7-142.

[66] Steinmann IC, Pflüger V, Schaffner F, Mathis A, Kaufmann C. Evaluation of matrix-assisted laser desorption/ionization time of flight mass spectrometry for the identification of ceratopogonid and culicid larvae. Parasitology 2013;140:318−27. Available from: https://doi.org/10.1017/S0031182012001618.

[67] Kauffmann C, Schaffner F, Ziegler D, Pflüger V, Mathis A. Identification of field-caught *Culicoides* biting midges using matrix-assisted laser desorption/ionization time of flight mass spectrometry. Parasitology 2012;139:248−58. Available from: https://doi.org/10.1017/S0031182011001764.

[68] Hoppenheit A, Murugaiyan J, Bauer B, Steuber S, Clausen PH, Roesler U. Identification of tsetse (*Glossina* spp.) using matrix-assisted laser desorption/ionisation time of flight mass spectrometry. PLoS Negl Trop Dis 2013;7:e2305. Available from: https://doi.org/10.1371/journal.pntd.0002305.

[69] Karger A, Kampen H, Bettin B, Dautel H, Ziller M, Hoffmann B, et al. Species determination and characterization of developmental stages of ticks by whole-animal matrix-assisted laser desorption/ionization mass spectrometry. Ticks Tick Borne Dis 2012;3:78−89. Available from: https://doi.org/10.1016/j.ttbdis.2011.11.002.

[70] Rothen J, Githaka N, Kanduma EG, Olds C, Pflüger V, Mwaura S, et al. Matrix-assisted laser desorption/ionization time of flight mass spectrometry for comprehensive indexing of East African ixodid tick species. Parasit Vectors 2016;9:151. Available from: https://doi.org/10.1186/s13071-016-1424-6.

[71] Yssouf A, Almeras L, Berenger JM, Laroche M, Raoult D, Parola P. Identification of tick species and disseminate pathogen using hemolymph by MALDI-TOF MS. Ticks Tick Borne Dis 2015;6:579−86. Available from: https://doi.org/10.1016/j.ttbdis.2015.04.013.

[72] Yssouf A, Almeras L, Terras J, Socolovschi C, Raoult D, Parola P. Detection of *Rickettsia* spp in ticks by MALDI-TOF MS. PLoS Negl Trop Dis 2015;9:e0003473. Available from: https://doi.org/10.1371/journal.pntd.0003473.

[73] Yssouf A, Flaudrops C, Drali R, Kernif T, Socolovschi C, Berenger JM, et al. Matrix-assisted laser desorption ionization-time of flight mass spectrometry for rapid identification of tick vectors. J Clin Microbiol 2013;51:522−8. Available from: https://doi.org/10.1128/JCM.02665-12.

[74] Dvořák V, Halada P, Hlaváčkova K, Dokianakis E, Antoniou M, Volf P. Identification of phlebotomine sand flies (*Diptera: Psychodidae*) by matrix-assisted laser desorption/ionization time of flight mass spectrometry. Parasit Vectors 2014;7:21. Available from: https://doi.org/10.1186/1756-3305-7-21.

[75] Mathis A, Depaquit J, Dvořák V, Tuten H, Bañuls AL, Halada P, et al. Identification of phlebotomine sand flies using one MALDI-TOF MS reference database and two mass spectrometer systems. Parasit Vectors 2015;8:266. Available from: https://doi.org/10.1186/s13071-015-0878-2.

[76] Yssouf A, Socolovschi C, Leulmi H, Kernif T, Bitam I, Audoly G, et al. Identification of flea species using MALDI-TOF/MS. Comp Immunol Microbiol Infect Dis 2014;37:153−7. Available from: https://doi.org/10.1016/j.cimid.2014.05.002.

[77] Muller P, Pfluger V, Wittwer M, Ziegler D, Chandre F, Simard F, et al. Identification of cryptic *Anopheles* mosquito species by molecular protein profiling. PLoS One 2013;8:e57486. Available from: https://doi.org/10.1371/journal.pone.0057486.

[78] Yssouf A, Parola P, Lindstrom A, Lilja T, L'Ambert G, Bondesson U, et al. Identification of European mosquito species by MALDI-TOF MS. Parasitol Res 2014;113:2375−8. Available from: https://doi.org/10.1007/s00436-014-3876-y.

[79] Murugaiyan J, Roesler U. MALDI-TOF MS profiling-advances in species identification of pests, parasites, and vectors. Front Cell Infect Microbiol 2017;7. Available from: https://doi.org/10.3389/fcimb.2017.00184.

[80] Ruhaak LR, van der Burgt YE, Cobbaert CM. Prospective applications of ultrahigh resolution proteomics in clinical mass spectrometry. Expert Rev Proteomics 2016;13:1063−71. Available from: https://doi.org/10.1080/14789450.2016.1253477.

Future Applications of MALDI-TOF Mass Spectrometry in Clinical Microbiology

Maria Ercibengoa[1] and Marta Alonso[1,2]
[1]CIBER Enfermedades Respiratorias-CIBERES, Madrid, Spain
[2]Servicio de Microbiología, Hospital Universitario Donostia, Donostia, Spain

17.1 INTRODUCTION

Currently, matrix-assisted laser desorption ionization—time of flight mass spectrometry (MALDI-TOF MS) is considered the best choice for microbial identification laboratories around the world, even though it has only existed for a decade. Among its functions is the identification of Gram-positive, Gram negative, aerobic, or anaerobic bacteria, as well as mycobacteria, yeast, or filamentous fungi cultivated on agar plates or liquid media.

MALD-TOF MS has also proved to be a reliable tool for other applications, such as antibiotic and antimycotic resistance testing in addition to for some typing applications. However, these procedures (antibiotic and antimycotic resistance and microorganism typing) are in their infancy and their usage is currently restricted to specialized laboratories

The future development of MALDI-TOF MS for clinically relevant microorganism identification will rely on the development of more machines that greatly reduce the amount of hands-on time for sample preparation. Even more critical to the development of MALDI-TOF MS will be the improvement of software and fully integrated databases within the device to digitalize the identification of microorganisms, thus reducing the need for skilled data analysis.

On the other hand, new fields of application for MALDI-TOF MS, such as imaging mass spectrometry (IMS) could be implemented in the near future. IMS allows the analysis of molecules directly from tissue sections and has been applied to address questions about skin pathophysiology, drug

The Use of Mass Spectrometry Technology (MALDI-TOF) in Clinical Microbiology.
DOI: https://doi.org/10.1016/B978-0-12-814451-0.00017-4

absorption, and metabolism. The development of this technology would be a challenge for clinical microbiology and pathologists to accelerate diagnosis when an infection is suspected and could potentially have a great impact on patient management [1,2].

17.2 LABORATORY AUTOMATION

Advancing automation in laboratories was initiated as a response to excessive workloads due to increases in medical examination request [3]. The automation process in clinical microbiology laboratories with the exception of serological testing has occurred at slower rate than in other laboratory areas mainly due to their technical complexity [4,5].

An important part of the preanalytical stage, and one that is extremely time-consuming, is the inoculation of samples on blood agar plates, which consumes more than 25% of the personnel's time who are responsible for the reception and initial processing of the samples [5].

There is at present equipment with varying degrees of automation: Total Laboratory Automation system from BD Kiestra (Drachten, The Netherlands), WASP (Walk Away Specimen Processor) of Copan Diagnostics, InoqulA (Brescia, Italy), and PREVI Isola of BioMérieux.

These systems allow the procedures to be reproduced more easily than manual inoculation on agar plates; the number of colonies of isolates is higher and manual workload is decreased [5−8].

However, the implementation of MALDI-TOF MS in the automation of microbiology laboratory is still in its infancy. Identification using MALDI-TOF MS continues to be performed manually (Fig. 17.1).

Manual colony picking is a major limitation of the throughput of MS identification and causes major errors due to inversions, which are estimated to occur at a rate of up to 3% [9].

As MALDI-TOF MS is currently used to identify more than 95% of all isolates in laboratories, automated systems including automated colony picking should be implemented. Such systems will undoubtedly have a major positive impact on the laboratory workload.

MALDI COLONYST [10] is the latest generation intelligent robot optimized for colony picking and complete MALDI target preparation either with Bruker MALDI Biotyper or with BioMérieux VITEK-MS. MALDI COLONYST can be connected to the FEEDER, a robotic arm for automated Petri dish handling. It can also be fully integrated into

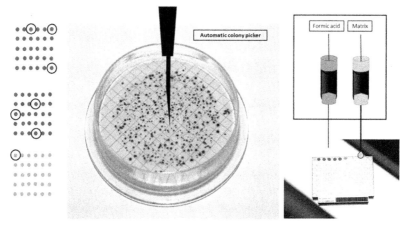

Figure 17.1 MALDI-TOF MS target plate automatic preparation.

laboratory information systems. In addition, it can export results and help to improve the laboratory workflow. This robot is the world's first complete device to fully automate the MALDI-TOF MS target plate preparation.

17.3 IMAGING MASS SPECTROMETRY

IMS is a new imaging technology based on MS that has existed for more than 20 years. This tool allows direct visualization of the spatial distribution of biomolecules in biological sections.

In brief, mass spectrometry is an analytical chemistry technique that identifies ionized molecules based on their mass-to-charge ratio (m/z) [11−13].

A mass spectrometer is composed of three main parts: an ionization source, an analyzer, and a detector. Both the ionization source and analyzer can differ in varying maker and models; however, the basic concept remains unchanged [14].

The raw data generated by spectrometer creates a unique fingerprint, known as mass spectrum. The analyte is a relevant biomolecule (lipids, proteins, and polysaccharides) and will be detected as a gas-phase ion [15,16].

There is great potential for IMS technology in the study of microbial systems. Many microorganisms lie in multicellular and multispecies communities adhered to the surface of the host area, such as biofilms and motile colonies where the bacteria work together to take advantage of the surrounding nutrients, fend off hostile organisms and with stand adverse environmental conditions [17].

These processes, like many others that are essential for the survival of the microbes are regulated by the production and usage of a great variety of chemical products.

IMS technology is well-positioned to study a wide variety of chemical interactions [18], including those which occur inside single-species microbial communities, between cohabitating microbes, and between microbes and their hosts (Fig. 17.2).

It is expected that many research areas will benefit from IMS technology in the near future, having a role in microbiology research and even diagnostic of metabolites without the need for any labeling at all. The capacity for the profiling of molecules directly from tissue samples by MALDI-TOF MS would open an attractive investigation field in different areas such as nanomedicine, pharmacokinetics, proteome imaging, and many others [19].

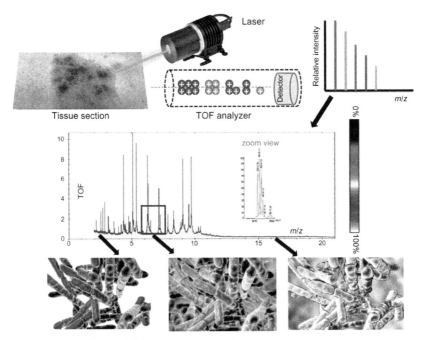

Figure 17.2 MALDI IMS Schematic representation.

17.4 IDENTIFYING BIOMARKERS IN CLINICAL SAMPLES FOR INFECTIOUS DISEASES DIAGNOSIS

MALDI-TOF MS plays an important role as proteomic research tool and offers the opportunity to discover biomarkers in clinical samples to improve the diagnose and also to contribute in the research of the pathogenesis of infectious diseases. Nowadays, MALDI-TOF MS technology is most widely used for discovering potential cancer biomarkers. MALDI-TOF MS presents advantages as it is an easy to use technique with high automation and throughput potential, and good sensitivity. The research of biomarkers for infectious disease diagnosis is a very challenging field. Samples as serum/plasma or urine would be the ideal samples to test as they are easy to obtain and adaptable to an automatized laboratory.

If serum or plasma is used, the sample must go through an enrichment process to avoid proteins that can be found in high concentration as albumin, immunoglobulins, or transferrin, which are estimated to represent the 90% of proteins in serum.

A technique derived from MALDI-TOF MS, surface-enhanced laser desorption/ionization (SELDI)−TOF is used in the research of new biomarkers for cancer and autoimmune diseases [20] but also in infectious diseases [21]. SELDI is typically used with TOF mass spectrometers for protein detection in tissue samples, blood, urine, or other clinical samples. To put it briefly, this technique consists in adhering the proteins to a surface ("protein chip") from which the MS is obtained. Once the MS is obtained, the identification of biomarkers is done by comparing the MS profile of patients with the disease under study with the control group (healthy individuals). In a subsequent step, the sensitivity and specificity of the selected biomarkers are tested.

Regarding infectious diseases research to SELDI-TOF and MALDI-TOF MS technique, specific biomarkers related to HIV1-dementia [22,23] and to predict the evolution of patients with chronic hepatitis C virus are found [24,25].

With regard to studies focused on clinical diagnosis, in the study of Zhang et al., the serum samples were enriched with weak cationic magnetic beads and then analyzed with MALDI-TOF MS. The researchers were able to identify specifics protein peaks with high specificity and sensitivity to distinguish intestinal tuberculosis from healthy patients and also from patients with Crohn's disease [26].

There are several studies in the area of clinical parasitology that have discovered biomarkers (host protein or parasite proteins) useful for the diagnoses and also for the understanding of the pathogenesis of these diseases.

Several biomarkers for the diagnoses of Chagas disease using SELDI-TOF technology have been described, and what is more, these biomarkers proved to be useful in the monitoring of the treatment with Nifurtimox [27,28]. This technology has also been used to detect a new biomarker for *Plasmodium falciparum* infection. In the future this biomarker could be used to develop a new rapid diagnosis technique that improves previous ones [29].

There are two interesting studies performed on animals which investigate the presence of biomarkers for the clinical diagnosis of infectious diseases caused by parasites. In an animal model with mice infected with *Schistosoma mansoni* the protein profile was notably different in the infected group vs control group. Besides, the protein profiles were different depending on the stages of the disease (early, acute of chronic) [30].

In another study performed on dogs, the serum profile of animals infected with Babesia was different from controls. There was a specific protein fraction only present in infected dogs. These proteins could be used as infection markers. The test would be easy and not expensive [31].

Another procedure in the research for new biomarkers is based in immuno-MALDI-TOF MS: immunoafffinity enrichment combined with MALDI-TOF MS. The combination of immunological methods with MALDI-TOF MS represents a significant improvement in the identification of new biomarkers. The traditional immunological methods such as enzyme-linked immunosorbent assay or radio immunoassay do not provide molecular information. Besides, immuno-MALDI-TOF MS, it is a more specific technique because the antibody is only used for antigen capturing, whereas traditional techniques are based in two-step procedures: one antibody is used to capture the target molecules and the second for the signal generation [32]. This methodology has been used also for the quantification of the targets, as for example, hypertension biomarkers [33], or more recently, in cancer research protein quantification [34].

Regarding infectious diseases, a potential clinical application of immuno-MALDI-TOF MS could help to improve the diagnoses *S. mansoni*−caused disease. Robijn et al. [35] described glycans in the urine of infected individuals that could be used as diagnostic marker. Further studies are still required to see if there is a correlation between the quantity of the glycan in urine and the egg load in liver.

The studies carried out so far lead us to be hopeful that in the future certain biomarkers detected with MALDI-TOF MS in serum/plasma would be useful to solve microbiology—related problems, e.g., toxin quantification. Finally, it also would be worth to explore if MALDI-TOF MS could identify new markers which would improve the early recognition of a patient with sepsis by Gram-positive, Gram-negative, aerobic or anaerobic bacteria, as well as mycobacteria, yeast, or filamentous fungi.

REFERENCES

[1] Chughtai K, Heeren R. Mass spectrometric imaging for biomedical tissue analysis. Chem Rev 2010;110(5):3237—77.
[2] Moore J, Caprioli R, Skaar E. Advanced mass spectrometry technologies for the study of microbial pathogenesis. Curr Opin Microbiol 2014;19:45—51.
[3] Young S. Laboratory automation: smart strategies and practical applications. Clin Chem 2000;46(5):740—5.
[4] Dumitrescu O, Dauwalder O, Lina G. Present and future automation in bacteriology. Clin Microbiol Infect 2011;17(5):649—50.
[5] Bourbeau P, Swartz B. First evaluation of the WASP, a new automated microbiology plating instrument. J Clin Microbiol 2009;47(4):1101—6.
[6] Glasson J, Guthrie L, Nielsen D, Bethell F. Evaluation of an automated instrument for inoculating and spreading samples onto agar plates. J Clin Microbiol 2008;46 (4):1281—4.
[7] Mischnik A, Mieth M, Busch C, Hofer S, Zimmermann S. First evaluation of automated specimen inoculation for wound swab samples by use of the Previ Isola System compared to manual inoculation in a routine laboratory: finding a cost-effective and accurate approach. J Clin Microbiol 2012;50(8):2732—6.
[8] Croxatto A, Prod'hom G, Faverjon F, Rochais Y, Greub G. Laboratory automation in clinical bacteriology: what system to choose? Clin Microbiol Infect 2016;22 (3):217—35.
[9] Bizzini A, Durussel C, Bille J, Greub G, Prod'hom G. Performance of matrix-assisted laser desorption ionization-time of flight mass spectrometry for identification of bacterial strains routinely isolated in a clinical microbiology laboratory. J Clin Microbiol 2010;48(5):1549—54.
[10] Colony Picking Robot for MALDI TOF Systems—Maldi Colonyst [Internet]. Colonyst.com. 2017 (cited 13 November 2017). Available from: https://www.colonyst.com/.
[11] Vickerman J. Molecular imaging and depth profiling by mass spectrometry—SIMS, MALDI or DESI? Analyst 2011;136(11):2199.
[12] Triolo A, Altamura M, Cardinali F, Sisto A, Maggi C. Mass spectrometry and combinatorial chemistry: a short outline. J Mass Spectrom 2001;36(12):1249—59.
[13] Biemann K. Mass spectrometry. Ann Rev Biochem 1963;32(1):755—80.
[14] Vanlear G, Mclafferty F. Biochemical aspects of high-resolution mass spectrometry. Annu Rev Biochem 1969;38(1):289—322.
[15] Glish G, Vachet R. The basics of mass spectrometry in the twenty-first century. Nat Rev Drug Discov 2003;2(2):140—50.
[16] Aebersold R, Mann M. Mass spectrometry-based proteomics. Nature 2003;422 (6928):198—207.

[17] Walch A, Rauser S, Deininger S, Höfler H. MALDI imaging mass spectrometry for direct tissue analysis: a new frontier for molecular histology. Histochem Cell Biol 2008;130(3):421−34.

[18] Yang J, Phelan V, Simkovsky R, Watrous J, Trial R, Fleming T, et al. Primer on agar-based microbial imaging mass spectrometry. J Bacteriol 2012;194(22):6023−8.

[19] Bérdy J. Thoughts and facts about antibiotics: where we are now and where we are heading. J Antibiot 2012;65(8):385−95.

[20] Engwegen JY, Gast MC, Schellens JH, Beijnen JH. Clinical proteomics: searching for better tumour markers with SELDI-TOF mass spectrometry. Trends Pharmacol Sci 2006;27(5):251−9.

[21] Ndao M, Rainczuk A, Rioux MC, Spithill TW, Ward BJ. Is SELDI-TOF a valid tool for diagnostic biomarkers? Trends Parasitol 2010;26(12):561−7.

[22] Luo X, Carlson KA, Wojna V, Mayo R, Biskup TM, Stoner J, et al. Macrophage proteomic fingerprinting predicts HIV-1-associated cognitive impairment. Neurology 2003;60(12):1931−7.

[23] Wojna V, Carlson KA, Luo X, Mayo R, Meléndez LM, Kraiselburd E, et al. Proteomic fingerprinting of human immunodeficiency virus type 1-associated dementia from patient monocyte-derived macrophages: a case study. J Neurovirol 2004;10(Suppl 1):74−81.

[24] Schwegler EE, Cazares L, Steel LF, Adam BL, Johnson DA, Semmes OJ, et al. SELDI-TOF MS profiling of serum for detection of the progression of chronic hepatitis C to hepatocellular carcinoma. Hepatology 2005;41(3):634−42.

[25] Göbel T, Vorderwülbecke S, Hauck K, Fey H, Häussinger D, Erhardt A. New multi protein patterns differentiate liver fibrosis stages and hepatocellular carcinoma in chronic hepatitis C serum samples. World J Gastroenterol 2006;12(47):7604−12.

[26] Zhang F, Xu C, Ning L, Hu F, Shan G, Chen H, et al. Exploration of serum proteomic profiling and diagnostic model that differentiate Crohn's disease and intestinal tuberculosis. PLoS One 2016;11(12):e0167109.

[27] Ndao M, Spithill TW, Caffrey R, Li H, Podust VN, Perichon R, et al. Identification of novel diagnostic serum biomarkers for Chagas' disease in asymptomatic subjects by mass spectrometric profiling. J Clin Microbiol 2010;48(4):1139−49.

[28] Santamaria C, Chatelain E, Jackson Y, Miao Q, Ward BJ, Chappuis F, et al. Serum biomarkers predictive of cure in Chagas disease patients after Nifurtimox treatment. BMC Infect Dis 2014;14:302.

[29] Thézénas ML, Huang H, Njie M, Ramaprasad A, Nwakanma DC, Fischer R, et al. PfHPRT: a new biomarker candidate of acute Plasmodium falciparum infection. J Proteome Res 2013;12(3):1211−22.

[30] Kardoush MI, Ward BJ, Ndao M. Identification of candidate serum biomarkers for Schistosoma mansoni infected mice using multiple proteomic platforms. PLoS One 2016;11(5):e0154465.

[31] Adaszek Ł, Banach T, Bartnicki M, Winiarczyk D, Łyp P, Winiarczyk S. Application the mass spectrometry MALDI-TOF technique for detection of Babesia canis canis infection in dogs. Parasitol Res 2014;113(11):4293−5.

[32] Sparbier K, Wenzel T, Dihazi H, Blaschke S, Müller GA, Deelder A, et al. Immuno-MALDI-TOF MS: new perspectives for clinical applications of mass spectrometry. Proteomics 2009;9(6):1442−50.

[33] Camenzind AG, van der Gugten JG, Popp R, Holmes DT, Borchers CH. Development and evaluation of an immuno-MALDI (iMALDI) assay for angiotensin I and the diagnosis of secondary hypertension. Clin Proteomics 2013;10(1):20.

[34] Popp R, Li H, LeBlanc A, Mohammed Y, Aguilar-Mahecha A, Chambers AG, et al. Immuno-matrix-assisted laser desorption/ionization assays for quantifying AKT1 and AKT2 in breast and colorectal cancer cell lines and tumors. Anal Chem 2017;89 (19):10592−600.

[35] Robijn ML, Koeleman CA, Hokke CH, Deelder AM. Schistosoma mansoni eggs excrete specific free oligosaccharides that are detectable in the urine of the human host. Mol Biochem Parasitol 2007;151(2):162−72.

FURTHER READING

Harkewicz R, Dennis E. Applications of mass spectrometry to lipids and membranes. Annu Rev Biochem 2011;80(1):301−25.

INDEX

Note: Page numbers followed by "*f*" and "*t*" refer to figures and tables, respectively.

Printed in the United States
By Bookmasters